GOOD GERMS, BAD GERMS

GOOD GERMS,
BAD GERMS

HEALTH AND SURVIVAL IN

A BACTERIAL WORLD

JESSICA SNYDER SACHS

HILL AND WANG

A DIVISION OF FARRAR, STRAUS AND GIROUX

NEW YORK

Hill and Wang
A division of Farrar, Straus and Giroux
19 Union Square West, New York 10003

Distributed in Canada by Douglas & McIntyre Ltd.
Printed in the United States of America
First edition, 2007

Grateful acknowledgment is made for permission to reprint the following previously
published material: pp. 43–44, excerpt from "A New Year Greeting" by W. H. Auden,
copyright © 1969 by W. H. Auden. Used by permission of Random House, Inc.

Photograph on dedication page by Dick Hostrander, used with permission from
Lycoming College.

Library of Congress Cataloging-in-Publication Data
Sachs, Jessica Snyder.
    Good germs, bad germs : health and survival in a bacterial world / Jessica Snyder
Sachs. — 1st ed.
        p.   cm.
    Includes index.
    ISBN-13: 978-0-8090-5063-5 (hardcover : alk. paper)
    ISBN-10: 0-8090-5063-3 (hardcover : alk. paper)
        1. Microbiology.   2. Medical microbiology.   3. Bacteriology.
    4. Communicable diseases.   I. Title
        [DNLM: 1. Bacteria   2. Anti-Bacterial Agents.   3. Drug Resistance,
Microbial.   4. Immunity, Natural.   5. Microbial Viability.   6. Microbiology—
trends.   QW 50 S121g 2007]
    QR57.S2387 2007
    616.9'041—dc22

                                                                    2007008526

Designed by Jonathan D. Lippincott

www.fsgbooks.com

10   9   8   7   6   5   4   3   2   1

This book is dedicated to Ricky Lannetti (1982–2003)

CONTENTS

SEVEN KEY TERMS AND CONVENTIONS

antibiotic In this book, I use the term antibiotic as a catchall for all bacteria-killing drugs. Those who split hairs over such things would use the word "microbicide" to refer to synthetic drugs used to kill bacteria and "antibiotic" for bacteria-killing drugs produced by living organisms such as fungi and soil bacteria.

archaea Bacteria-like organisms that split off from true bacteria early in the evolution of life on Earth. They are best known for surviving at extreme temperatures and producing methane. None are known to cause disease, but several make their homes in our mouths and intestines. For simplicity's sake, when I refer to bacteria as a group, I include the archaea.

bacteria The subject of this book: simple one-celled organisms that lack a nucleus and membrane-bound cell structures. This type of cell is also called *prokaryotic*, to distinguish it from the larger and more compartmentalized *eukaryotic* cells of so-called higher life.

germ A catchall term for infectious microbes and viruses.

microbes Microscopic organisms, a group that includes prokaryotic bacteria as well as eukaryotic parasites such as fungi (vaginal "yeasts," ringworm, and others) and protists (malaria, giardia, and kin). Some scientists include viruses in this group. I do not, as viruses are not organisms in the sense of being alive.

microflora Not tiny flowers, but the resident microbes of the human body. Bacteria predominate our microflora, but the category also includes a scattering of fungi. Protists, while not a part of a healthy microflora, can and do take up residence as parasites.

virus An infectious particle consisting of genetic material (either DNA or RNA) encapsulated in a protein coat. Not technically alive, viruses can cause disease when they are taken up by cells and their genetic material redirects those cells into making more viruses. Bacteriophages are viruses that infect bacteria, and so are an important part of this book.

A GOOD WAR GONE BAD

RICKY'S STORY

On the icy afternoon of December 6, 2003, Theresa Lannetti slipped into the home-team bleachers of Person Stadium, in the wooded college town of Williamsport, Pennsylvania. Like everyone around her, she had come to cheer the Lycoming College Warriors in their NCAA division quarterfinals against the Bridgewater Eagles. But for once Theresa did not scan the crowd for the other parents with whom she'd become friends over the previous three years. She understood that few would expect her to be there, and some might feel awkward that she was.

The previous night, in the critical-care unit of a local community hospital, Theresa had wrapped her arms around her son's college roommate, Sean Hennigar, telling him that he would score a touchdown for Ricky the next day. In the morning Theresa knew that she had to stay in Williamsport and lose herself in the familiar sounds of the game. She watched as Ricky's teammates jogged out from the locker room; she registered the clenched jaws and fists of his best friends; and then she glimpsed the 19s scrawled in eye black across their arms, sleeves, and pant legs. The cheerleaders rose to welcome the team—more 19s, written in jagged strips of athletic tape across the backs of their jackets and with blue greasepaint on puffy cheeks streaked with tears. Theresa's own eyes remained dry until she looked across the field to see yet another 19, some fifteen feet tall, shoveled into the snowy hillside alongside the visitor stands. As Theresa watched, the artist put down his shovel and flopped onto his back, adding one perfect snow angel and then another—two asterisks above her son's number. Perhaps they stood for the two team records that Ricky Lannetti had set that season: six pass receptions in a single game and seventy over the regular season. Overhead, the clouds began to part for the first time that week.

The first snowstorm of the winter had arrived the previous Tuesday, dusting the field as the Warriors finished afternoon practice. The forecast of weeklong blizzard conditions did not dampen the campus-wide excitement over the NCAA division quarterfinals, with expectations high that this would be the season Lycoming would advance to the semifinals, having fallen short the previous six years. Ricky had started coughing that morning, and near the end of the blustery cold practice, a wave of nausea forced him to sit out the final plays. When Theresa called from Philadelphia the next day, Ricky cut her short. "Mom, I can't talk right now. I don't feel so good," he said. "I'll be fine. It's just one of those twenty-four-hour things. I gotta throw up now, okay?"

Theresa told herself that Ricky was probably right—he'd be fine by the weekend. As always, she looked forward to watching her son play and didn't even mind the 180-mile trip from her row-house neighborhood in northeast Philly to Williamsport, the central Pennsylvania college town where Ricky was halfway through his senior year as a criminal justice major. While Theresa's marriage to Ricky's father had ended in 1991, their son's sports, especially football, kept family and mutual friends bound together around practice schedules, games, and postgame celebrations. In grade school Ricky had a reputation as a "mighty mite," always among the smallest on the field but also the hardest to catch or evade. Other kids his age could never understand how a guy who barely reached their shoulders could hit so damn hard or arc into the air, as if from nowhere, to snag a ball. By the time Ricky reached high school, Theresa was cheering more often than cringing when her son withstood tackles that should have brought down someone twice his size. She even begrudgingly accepted the fact that there was no way she could drag him out of a game, even when she knew he'd been hurt. In his four years playing for Lycoming, he had missed only one game, and that for a badly twisted ankle that he insisted was fine after a week.

But the days leading up to the team's quarterfinal game had Theresa worried. Ricky was still throwing up when she called back on Thursday. "You can't ignore this!" she insisted. "You have to have someone check this out." She phoned the head trainer, Frank Neu, who promised to take Ricky to see his wife, a family physician, to rule out anything more serious than the flu that had been downing students since the Thanksgiving break. That afternoon Stacey Neu listened to Ricky's lungs. They sounded clear. She took his temperature—slightly elevated. With his main symptoms being nausea, fatigue, and achiness, all signs pointed to the flu. Antibiotics wouldn't help, she explained, as they kill bacteria, not viruses.

The snow was still falling across Pennsylvania on Friday morning when Ricky's legs began to ache. His roommates, Sean Hennigar and Brian Conners, pushed sports drinks and water, figuring that Ricky was dehydrated from vomiting. That evening the snow slowed Theresa's drive out of Philly to a crawl. She wasn't yet halfway to Williamsport at 9:30 p.m. when Ricky called her cell phone to say he wanted to sleep. He'd see her in the morning. But Ricky didn't sleep. And despite himself, he moaned every time he rolled over in bed. "If I don't get outta here, you guys won't get any sleep before the game," he told his roommates at 4:00 a.m. He called and asked his mom to bring him to her hotel room.

Initially, Ricky rebuffed Theresa's suggestion that they go to the hospital. But when she heard him sound short of breath, she insisted. At 7:20 a.m. they pushed through the emergency room doors, and Ricky began vomiting blood. Within five minutes a half dozen doctors and nurses had oxygen flowing into Ricky's nostrils and an intravenous line infusing fluids into his left arm. They drew blood for laboratory tests and put him on a heart monitor.

The critical-care doctor in charge noted Ricky's rapid heart rate, low blood pressure, shortness of breath, and slightly elevated temperature. "Just achy all over," Ricky told him when the doctor asked about chest or abdominal pain. "And tired, really tired." Ricky's lungs still sounded clear. His nose and throat looked normal. The doctor asked when he last peed. "Sometime Thursday, I think," Ricky replied. Theresa saw the alarm in the face of the doctor, who immediately called for a catheter. As soon as the nurse inserted it, the attached Foley bag filled with cloudy brown urine. Ricky's kidneys were shutting down.

Though Ricky's symptoms still pointed to a viral illness—albeit an overwhelming one—the critical-care doctor decided that he could not afford to wait for the twelve-hour blood culture that could rule out a bacterial infection. He added two powerful antibiotics—cefepime and vancomycin—to the fluids flowing into Ricky's arm. Theresa, meanwhile, had reached Ricky's father en route to Williamsport for the afternoon game and diverted him to the hospital. When Rick Senior arrived just before 11:00 a.m., he was stunned by the sight of his son in a tangle of tubes and wires, lips and nose still flecked with dry blood.

Back at Lycoming College, NCAA officials were trodding over the snow-blanketed playing turf and kicking at the broad mound of ice that encased midfield. At 10:00 a.m. they had announced their decision to postpone the game to Sunday rather than risk player injury. Ricky's closest teammates headed to the hospital, where to their surprise the emergency room staff let them walk into the intensive care unit. Ricky seemed better. Irritated with the parade of doctors in and out of his room, he wanted the oxygen line removed from his nose, and he wanted to know when he could go home. "I heard the game was postponed," he told Sean. "That's great. Maybe I'll have a chance to play."

But just after 1:00 p.m. Ricky sank into unconsciousness, and alarms began beeping as the level of oxygen in his blood dropped dangerously low. A nurse herded the football players out of the room, as doctors snaked a tube down Ricky's throat so a mechanical ventilator could take over the work of his failing lungs. An echocardiogram showed that his heart was likewise weakening, and a call went out for a helicopter to fly Ricky to Philadelphia's Temple University Medical Center, where he could be put on a heart-assist machine. An infectious-disease specialist ordered additional antibiotics as a hedge against any conceivable type of bacterial infection. Something was destroying Ricky's organs, but exactly what or where it lurked in his body remained maddeningly unclear. Then came worse news: the helicopter pilot at Temple University Hospital had ruled it too dangerous to fly in the continuing snowstorm.

Ricky's heart stopped at 5:36 p.m. For forty minutes, nurses and doctors performed CPR as they waited for a cardiac surgeon to arrive. By 6:30, the surgeon had connected Ricky's arteries to a heart-lung bypass machine that circulated sixteen units of oxygenated blood through his body. But an hour later Ricky's heart remained still, his eyes fixed and dilated. The surgeon stepped out of the operating room to talk with Ricky's parents. At 7:36 p.m. he disconnected Ricky's machine.

Theresa asked the nurses to help her clean Ricky and move him back to his hospital room. She wanted the boys—now sobbing in the hallways—to see their friend once more before they left. She knew she had done the right thing when Ricky's father pointed out the smile on his son's face and linebacker Tim Schmidt added, "Damn, if that's not that smirk he gets when he's just pulled something over on us."

A few minutes later Ricky's parents learned the name of his killer. The blood culture the doctors had ordered that morning came back positive for methicillin-resistant *Staphylococcus aureus*, MRSA, a bug that could shrug off not only the methicillin family of antibiotics but also a half dozen others. Worse, this particular strain of MRSA—now known to specialists as USA300—also carried genes for an array of toxins, some of which triggered the deadly internal storm known as septic shock, with its signature symptoms of plummeting blood pressure, massive blood clotting, and organ failure.

The morning after Ricky's death, a muffled quiet saturated the Lycoming locker room even as the team began to gather for the coming game. Several exhausted players had been at the hospital through the night. But all agreed that Ricky would have been "royally pissed" if they did not play. As Ricky's devastated father let a family friend drive him home, Theresa took her seat in the stands. The game began fitfully, as the Warriors struggled with plays that had been designed around their star receiver. Early in the first quarter, quarterback Phil Mann hesitated in the huddle before calling, "Club right." The play had always meant Ricky, running like hell out of the offensive backfield in a narrow crossing pattern. This time it would be Sean, sprinting from behind the thirteen-yard line. When Sean turned, Mann's missile smacked hard into his hands. He bulldozed the last three yards to the touchdown. It would be the only score for Lycoming in its losing bid for the semifinals that day. But for that moment, high in the stands, Theresa found herself screaming with the crowd, shaking and crying as, out on the field, Sean Hennigar lifted the football high toward a now-blue sky.

DANIEL'S STORY

At age seven, Daniel tries to shrug it off. But it's kinda scary when classmates brag they're "gonna get him" with their peanut butter sandwiches. "It horrifies me," says his mother, Ann, a New York City radio producer.[1] "How can they remotely think this funny?" She remembers the November evening, five years ago, when she almost lost her beautiful dimple-cheeked boy. It began innocuously enough. Ann had just stepped onto the sidewalk in front of her Manhattan office and pulled out her cell phone to let the babysitter know she was running a little late. Daniel had just thrown up after eating an almond-butter sandwich, the sitter reported. "Okay, these things happen," Ann reassured her. "Just keep an eye on him." When Ann called from Penn Station fifteen minutes later, Daniel had diarrhea. When she called a third time from the train out of New York City, Daniel was struggling for breath. "I'm giving him his asthma nebulizer," the babysitter said. "Call nine-one-one. Right now," Ann insisted.

When Daniel's mother next saw him, on a gurney emerging from an ambulance at their local New Jersey hospital, large red welts covered his face. The paramedics had found Daniel in anaphylactic shock, his throat swelling shut and his blood pressure plummeting toward zero. They brought him back with injections of the powerful stimulant epinephrine and an inflammation-stopping steroid. That's when Daniel's parents learned that he had life-threatening food allergies—most severely to peanuts, a trace amount of which had likely cross-contaminated the almond butter he'd eaten.

"Looking back, he probably had allergies long before we realized," Ann says. When Daniel was two months old, he developed severe eczema, an allergic skin condition that left his cheeks pink and scaly and the inside of his elbows and backs of his knees raw and oozing. Shortly after Daniel's first birthday, he developed asthma, yet another condition often triggered by allergies, in this case to respiratory allergens such as dust mites and dander. By his second birthday fresh blood was appearing in his stools. It was just a little at first, and the pediatrician told Ann not to worry. Children often bled a little from a small fissure, or irritated crack, inside the rectum.

But over the next three years, not only did Daniel's allergies become obvious and life-threatening, the bleeding worsened, until he was passing more than twenty bloody stools a day with horrific cramping. A colonoscopy revealed open sores the length of Daniel's large intestine. He had ulcerative colitis, an inflammatory disorder in which the immune system mistakes either food or the intestine's normal resident bacteria for foreign invaders deserving a full-bore attack. At the start of second grade Daniel spent three weeks in the hospital on powerful anti-inflammatory drugs and went home with a long-term prescription for sulfa drugs to keep the inflammation from returning. But within a year, Daniel developed the telltale rashes that indicated he was becoming allergic to his medications. In the fall of 2006 his colitis returned, putting him at risk for a perforation of the colon—a condition as potentially deadly as a burst appendix.

"No child should have to go through this," says Ann. "Our challenge has been trying to figure out how to protect him while letting him live a normal life." Ann has had to decide, for example, whether to allow Daniel to eat lunch in the school cafeteria, his allergies being so severe that contact with something as small as a peanut crumb could trigger anaphylaxis. The compromise: her son sits at the end of a lunch table next to his best friend, whose mother knows not to pack peanut butter or anything else on Daniel's list of forbidden foods.

Not that Daniel is alone. "There are five kids in his grade who have EpiPens," says Ann of the prefilled prescription syringes that contain enough epinephrine to reverse a life-threatening allergic reaction. Daniel never leaves the house without two of them strapped to his waist in a special belt. Daniel's eczema, asthma, and respiratory allergies are even more commonplace among his classmates, says his school nurse. In all, she estimates that 40 percent of her students have some type of serious allergic disorder. "We're not talking about a little sneezing in hay fever season," she says. "These kids are sick. Many can't go outside for recess or field trips." After thirty years as a nurse, she has also been taken aback by a recent increase in inflammatory bowel disorders in her students—not only Daniel's ulcerative colitis, but even more commonly, Crohn's disease and irritable bowel syndrome, problems she'd never before seen in children.

REVENGE OF THE MICROBES?

On the surface, Ricky's and Daniel's stories may seem to have little in common. One involved a fatal, drug-resistant infection; the other, a trio of life-threatening inflammatory disorders. Yet both appear to stem from our modern-day war on germs. In the fifty years since antibiotics have come into widespread use, we have bred varying levels of drug resistance into every known kind of disease-causing bacterium. We've done this, in part, by carpet-bombing the body's normal resident microbes whenever we treat infections caused by the occasional invader. We've also been dousing our livestock with antibiotics, not only to cure their infections but also to spur their growth so they can be brought to market more quickly and cheaply. Some of their highly drug-resistant bacteria have likewise ended up in our bodies via the dinner plate.

In the process, we have run through the potency of more than a hundred different antibiotics, so that drug resistance is now a routine challenge of modern medicine, from the treatment of childhood ear infections to that of wholly unstoppable strains of tuberculosis. Like Ricky's killer, the USA300 staph strain now circulating throughout North America, some of these supergerms combine multidrug resistance with extreme virulence.

At the same time, immunologists and other medical experts have built a strong case that an unprecedented epidemic in inflammatory diseases stems from the second front in our long-standing war on germs, sanitation—from modern sewage systems and chlorinated water to refrigeration and food processing. Rare to nonexistent before the 1800s, these disorders include allergies and allergy-triggered asthma (which result when our immune cells react to harmless substances in our food and environment), as well as inflammatory bowel diseases like Daniel's. Also on a dramatic increase in the developed world are dozens of autoimmune diseases such as type 1 diabetes, lupus, multiple sclerosis, and rheumatoid arthritis, each of which results when the immune system mistakenly destroys healthy tissues. Tellingly, all these inflammatory disorders remain rare in regions where people still live in close contact with the soil, drink unfiltered water, and eat minimally processed and often crudely stored food.

So what is it about some bacteria that is so beneficial to us? After a century of focusing exclusively on disease-causing microbes, medical researchers are just starting to investigate. Environmental scientists, by contrast, have long understood that our very atmosphere and the recycling of life's nutrients depend on the bacterial kingdom's ceaseless activities. As the evolutionary microbiologist Lynn Margulis has quipped, "The moon is what our planet would look like without microbes." Less appreciated has been the reality that bacteria dominate the landscape of the human body as well, if sheer numbers be the gauge. These tiny one-celled organisms—which carpet our skin and line our digestive and upper respiratory tracts—outnumber our (admittedly larger) human cells by a factor of ten to one. This turns out to be a very good thing, for the body's resident microflora form a kind of protective mulch that has always been our best defense against infectious disease. In addition, the human immune system evolved to tolerate the ocean of largely harmless bacteria that flow through our bodies in our food, water, and air. The absence of their constant, reassuring touch appears to leave the immune system on hair trigger, with a nasty tendency to shoot up the neighborhood.

Still, no one yearns for the days before antibiotics, when doctors could do little more for their feverish patients than wait to see if they survived the night. Nor would any reasonable person propose that we trade modern sanitation for the epidemics of cholera, dysentery, typhoid fever, and bubonic plague that began decimating populations with the advent of civilization some five thousand years ago. Instead, a scientific consensus is growing: that only by understanding the symbiotic aspects of our long-standing relationship with microbes can we find lasting solutions to infectious disease and, at the same time, rectify the imbalances that have produced a modern epidemic of allergies, autoimmune disorders, and other inflammatory diseases.

Clearly, mounting a direct assault on the bacterial kingdom has always been foolhardy, given how rapidly these organisms can evolve around any biochemical weapon we throw at them. Rather than escalate an arms race we can never win, many scientists are now looking for better approaches to the problem, including

- teaching physicians to choose the right "sniper bullet" of an antibiotic instead of the popular "big guns" that tend to destroy the body's protective bacteria along with the bad;
- exploring new avenues of drug development aimed at loosening a destructive microbe's hold on the body without breeding drug resistance;
- studying why two people can harbor the same disease-causing microbes in their bodies but only one gets sick. In an age when we can compare one person's genome with the next

person's, the potential exists to use this understanding to develop therapies that leave the microorganism in place and instead fix the patient; and

- listening in on the intimate biochemical chatter that takes place between the body's resident microbes and its human cells in an effort to understand further why a body filled with the right kind of germs is a very healthy thing.

We have even begun domesticating our problem microbes, much as our ancestors once turned wolves into animals that guarded, rather than preyed upon, their sheep. Already this futuristic approach has enjoyed some early successes: a "probiotic" nasal spray imbued with beneficial bacteria that help prevent chronic childhood ear infections; a bioengineered strain of mouth bacteria that prevent rather than cause cavities; and a so-called Dirt Vaccine that appears to ease a range of chronic inflammatory disorders and also jolts the immune system into cancer-fighting mode. Some scientists are even dreaming about "probiotic" cleaning products—each detergent, cleanser, or air spray formulated with its own patented mix of protective and health-enhancing microbes.

But we stand at a crossroads. Proposals for peaceful coexistence with microorganisms can sound naïve in the face of the desperate need for new antibiotics to counter the growing threat of deadly, drug-resistant infections. In addition, a new generation of microbe hunters has begun to use genetic analysis, or DNA fingerprinting, to comb the body for infections we never knew we had. Their claim—that hidden infections lie behind such common ills as arthritis, heart disease, and Alzheimer's—may in fact be true, at least in part. The danger lies in the possibility that discovering such infections will set off a call to arms that brings about a historic increase in antibiotic use before we have applied the hard-won lessons of our first Hundred Years War on germs.

Here then is the story of our coevolution with the bacteria that imbue our lives, our profoundly altered relationship with their kind, and the possibility that we may yet achieve lasting health and continued survival in what has always been—and no doubt always will be—a microbial world.

PART I

THE WAR ON GERMS

Came 1910, and that was Paul Ehrlich's year. One day, that year, he walked into the scientific congress at Koenigsberg, and there was applause. It was frantic, it was long, you would think they were never going to let Paul Ehrlich say his say. He told of how the magic bullet had been found at last.

—PAUL DE KRUIF, *The Microbe Hunters*, 1926

FROM MIASMAS TO MICROBES

As scientific geniuses go, Girolamo Fracastoro of Verona was an accommodating sort. If a medical theory didn't suit his politics or his patrons, the famed Renaissance physician could not only rationalize it away but do so in Latin hexameter. A clinician by necessity, a poet and scholar by inclination, he was justly proud of his poetical treatise, *Syphilis sive morbus gallicus,* or "Syphilis, the French Disease," published to great acclaim in 1530.[1] Though most of Europe blamed Columbus and his Spanish sailors for bringing the "Great Pox" back from the New World, the title of Fracastoro's epic poem conveniently shifted the blame to the current enemies of Verona's Hapsburg occupiers (the Hapsburgs having aligned with Spain through marriage). Even Fracastoro's coining of the disease's name sprang from political gymnastics. For into his poem he wove the tale of Syphilis the shepherd, a resident of Atlantis who is instantly cured with a sip of guaiacum, the supposedly medicinal resin of a New World tree whose bark the Hapsburgs were importing en masse.

More profoundly, the poem's disease terminology marked a turning point in Western medicine, one that set it on a course toward direct warfare on the microbial world. Fracastoro was the first to set down in words the idea that germs—invisible contagious elements—exist in some kind of physical form. He wrote,

> And such as Nature must with pangs bring forth,
> Were violent and various Seeds united,
> Break slowly from the Bosom of the Night
> Long in the Womb of Fate the Embryo's worn
> Whole Ages pass before the Monster's born.[2]

Fracastoro's use of the word "seed" reflected the emerging beliefs of a small number of his contemporaries. These enlightened men and women had begun challenging the Hippocratic mindset of the previous millennium, which held that all disease derived from imbalances in the body's three "humors"—blood, phlegm, and bile.

Admittedly, the prototypic germs portrayed in Fracastoro's poetry were far from the microorganisms we know today. He used the terms "seeds" and "germs" to refer to elements more akin to atoms than to living organisms such as the spiral-shaped syphilis bacterium, *Treponema pallidum*, which was then wriggling its way into European genitalia from the Papal Court to the Scottish Highlands. Fracastoro also attributed an alignment of the planets as the "powerful cause" that drove these invisible seeds out of the "Sea and Earth and into the Aire," where they simultaneously infected the entire planet—conveniently explaining why Columbus and his sailors became infected in the New World around the same time as the disease erupted in the Old.

Certainly the idea of contagion—that is, the person-to-person transmission of disease—was not new to Renaissance Europe. Towns had begun quarantining plague victims, to little advantage, a century earlier. And the physicians of Egypt, India, and China had been grappling with the undeniable infectiousness of smallpox, a viral disease, since at least 3700 B.C. Indeed, all the world's early civilizations knew contagion, for it was the crowding of civilization that fueled it.

Over the millions of years that humans and their ancestors lived as hunter-gatherers, their populations remained too small to allow a deadly infection to last long, or to travel far before either killing everyone off or engendering population-wide immunity in the survivors—in either case, bringing extinction to the infectious organism. Exceptions included microbes that used humans as secondary homes while residing primarily in animals such as insects (which they did not harm). Of these, the mosquito-borne malaria parasite, a protozoan microbe,[3] may be the oldest and deadliest.

When tribes of *Homo sapiens* first hiked north out of Africa some thirty-five thousand years ago, they largely escaped their tropical parasites and enjoyed a prolonged era of robust health. The cave paintings of Europe's prehistoric wanderers show no hint of epidemics; nor do the legends of the New World's first nations. This is not to say that hunter-gatherers enjoyed an idyllic existence. Starvation and injury made for a short and brutal life span, but it was nonetheless a life span largely free of infectious disease.

Permanent settlements brought the stability of yearly harvests and domesticated livestock and the safety of fortress walls. The trade-off: crowding and water contamination. With civilization, well-behaved microbes abruptly lost their near-monopoly over the human body and a new microbial lifestyle arose—one in which virulence paid off, given that deadly bacteria could count on the coughing and flux of the dying to contaminate the air and water shared by thousands living in close quarters.

Epidemiologists calculate that it takes a population of around half a million to perpetuate an infectious disease: in other words, to allow a disease-causing microbe to continue jumping from host to host ahead of death or complete recovery. Not coincidentally, the first recorded mention of "pestilence" dates to the first civilization to reach that population benchmark: Sumeria, a string of a dozen merchant-trader cities on the delta of the Tigris and Euphrates rivers in what is now southeastern Iraq. The four-thousand-year-old Epic of Gilgamesh, Sumeria's version of the great flood story, references the ravages of Ura, the plague demon, as being preferable to the terrible flood, for at least Ura would have left a few survivors to repopulate.[4]

Like people in most ancient cultures, the Sumerians attributed the advent of plagues to angry gods and demons. So cures consisted of attempts to appease the heavens. As early medical traditions developed, the focus shifted away from finding the cause to finding symptomatic relief. The Hippocratic tradition, for example, relied on assessing imbalances in a patient's inner energies (as evidenced by fever, pus, and other symptoms) and then bleeding, purging, or sweating the person back into "equilibrium." It mattered not what caused a particular imbalance (too much bile, not enough blood, etc.), for the cure would be the same.

Syphilis may have been the mighty slap that convinced Europeans to look for a physical, albeit invisible, cause for infection. The magnitude of the syphilitic scourge of the sixteenth century can be difficult to appreciate today, for the disease has become far less virulent as man and microbe have coevolved over the centuries. Today only one in nine people infected with the syphilis bacterium develops symptoms obvious enough to send him or her to the doctor. Contrast this to the gaping sores, unrelenting pain, blindness, madness, and death so commonly portrayed in Renaissance literature and art.

The tremendous renown of Fracastoro's *Syphilis*, still the world's most famous medical poem, brought him the kind of fame and enduring patronage that allowed him to retire both from politicking and from clinical practice. With this newfound freedom, he produced the less poetic but far more scientifically important treatise *De contagione*, published in 1546.[5] It lays out the revolutionary elements of his more fully developed germ theory of disease:

- that infectious disease *always* spreads via invisible contagious seeds—*seminaria contagionis*;
- that it does so in three ways: *contactu afficit* (by direct contact), *fomite afficit* (by contact with contaminated objects), and *distans fit* ("at a distance," as in through the air);
- that germs have distinct identities: fevers are not all alike, nor can the germ that causes syphilis one year turn around and cause leprosy the next; and
- that different diseases will respond to different remedies, including various methods for directly attacking, or "burning out," the germs from the patient's body. In the case of syphilis, for example, Fracastoro mentions mercury, or quicksilver (a brutal remedy, it turns out, for it destroys not only the fragile syphilis spirochete but also massive numbers of human brain cells).[6]

For nearly four centuries, Fracastoro's category of "contagion at a distance" would linger as the vague concept of miasmas, or "poisonous aires." Belief in miasmas had become fully entrenched by the time the Black Death, or bubonic plague, showed its face in London during the broiling summer of 1665. To prevent the escape of these deadly "aires," communities nailed shut the doors of those who were stricken, and passersby kept their noses in "rings," or small bouquets, of posies, whose strong fragrance they hoped would deflect the toxic fog.[7]

The real carriers of the infectious organism—rats and their attendant fleas—had massively multiplied, thriving in the household waste that the working poor tossed from their windows for lack of a better option. By the end of the summer, more than thirty-one thousand people, or 15 percent of the city's population, were dead, and the king, his court, and most everyone with means had fled the city. Samuel Pepys, a naval secretary, described the desolation on the eve of his own departure, writing in his diary, "But now, how few people I see, and those walking like people that have taken leave of the world."[8]

Despite these horrific conditions, the same year brought the publication of the first of many scholarly books written by the "natural philosophers" of London's Royal Society. Robert Hooke filled his *Micrographia* with drawings of the fantastical structures that he had glimpsed beneath his magnifying lenses: the facets of a fly's eye, a louse clinging to a human hair, the individual "cells" (a term of Hooke's coinage) in a slice of cork.[9]

Neither plague nor war with England could keep copies of *Micrographia* out of the Netherlands, where the doe-eyed fabric merchant Antoni van Leeuwenhoek (pronounced *Lay-wen-hook*) fell under its spell.[10] So inspired, Leeuwenhoek taught himself lens grinding and constructed the first of hundreds of microscopes with which he would eventually discover the world of microscopic organisms. Leeuwenhoek's earliest discoveries included the giants of the microbial world: the one-celled algae and protozoas that floated in swamps near his home.[11] Yet for pure shock value, nothing came close to Leeuwenhoek's findings of 1683, when he turned his increasingly powerful microscopes on himself. In a letter dated September 17, 1683, he describes the hyperactive zoo that he found inside a mixture of his own tooth scum and spittle, enclosing a drawing that now ranks as an icon of microbiology:

I then most always saw, with great wonder, that in the said matter there were many very little living animalcules, very prettily a-moving. The biggest sort had the shape of Figure A. The biggest sort had a very strong and swift motion, and shot through the water (or spittle) like a pike does through water. These were most always few in number . . . The second sort had the shape of Fig. B. These oft-times spun round like a top, and every now and then took a course like that shown between C and D: and these were far more in number. To the third sort I could assign no figure, for at times they seemed to be oblong, while anon they looked perfectly round. These were so small that I could see them no bigger than Fig. E: yet therewithall they went ahead so nimbly, and hovered so together that you might imagine them to be a big swarm of gnats or flies, flying in and out among one another. These last seemed to me e'en as if there were several thousand of 'em in an amount of water or spittle no bigger than a sand-grain.[12]

Neither Leeuwenhoek nor his esteemed London correspondents viewed the fantastical "animalcules" he had discovered as anything but harmless, and for the most part they were. The middle-aged Dutchman's own remarkable health and vigor seemed testament enough to that. He even had all his own teeth. What's more, the idea that such fragile creatures could injure someone would have seemed laughable. As Leeuwenhoek himself observed, a mere sip of hot coffee or wine vinegar could make the animalcules on the surface of his teeth to "fall dead forthwith."[13] For the next century microbiology would remain a fringe science—one of interest only to the naturalists who had to describe and name the inhabitants of this lush new kingdom of life.

GERM THEORY REBORN

The late 1700s brought the opening of Europe's first maternity wards, or lying-in hospitals—a great development in public health designed to benefit the wealthy and poor alike. But the time proved to be especially deadly as well, with puerperal, or childbed, fever racing through the newly popular maternity wards and killing thousands. And no wonder, for doctors and midwives were constantly moving between the sick and the merely birthing, thrusting contaminated hands and instruments high into raw and torn birth canals and wombs. But the idea that medical workers might be spreading infection was shared by few and shunned by many. In fact, the controversial concept would destroy many careers.

The first was that of the Scottish surgeon Alexander Gordon. In his 1795 *Treatise on the Epidemic Puerperal Fever of Aberdeen*,[14] Gordon wrote,

> That the cause of this disease was a specific contagion, or infection, I have unquestionable proof . . . [for] this disease seized such women only, as were visited, or delivered, by a practitioner, or taken care of by a nurse, who had previously attended patients affected by the disease.

Gordon also noted the resemblance between the milky substance seen in the wombs of women dead of puerperal fever and the pussy discharge of erysipelas, or wound infections. He pointed out that "if in a dissection of a putrid [pus-containing] body, a surgeon scratches his finger, the part festers." Gordon's proposed cure for puerperal fever was all but medieval: bleeding off a pint and a half of blood from the woman, who in many cases had already bled copiously through childbirth. But his recipe for preventing its spread was spot-on:

> The patient's apparel and bedcloths ought either be burnt or thoroughly purified; and the nurses and physicians who have attended patients affected with the puerperal fever ought carefully to wash themselves and get their apparel properly fumigated [with smoking sulfur] before it be put on again.

Similar measures were already in use for stopping outbreaks of measles and smallpox, which were thought to arise from miasmas and then propagate from person to person.

Gordon would be the first of a string of physicians who offended far more people than they convinced with their polemics on hand-washing and sterilization. A half century later, a young Oliver Wendell Holmes tried unsuccessfully to hector doctors on the left side of the Atlantic into recognizing the infectious nature of puerperal fever. Dismissed as another crazy "contagionist" physician, he left medical practice four years later, in 1847, to teach at Harvard and earn international fame for his literary wit, even as the world forgot his medical insight.[15]

The same year that Holmes abandoned medical practice, the Hungarian physician Ignaz Semmelweis supplied clear proof of Gordon's and Holmes's contagionist theories. At Vienna's world-famous teaching hospital, the Allgemeine Krankenhaus, Semmelweis reversed skyrocketing rates of childbed fever by insisting that doctors and medical students scrub their hands with chloridated lime (powdered bleach) between performing autopsies and assisting women in labor. But in the process, the scowling and socially maladept Hungarian so insulted his Viennese colleagues with his implications of their slovenliness that they turned against him. Stripped of his research privileges at the Vienna Medical School in 1850, Semmelweis abruptly resigned and returned to his native Hungary, where he suffered a mental breakdown. With their would-be savior locked in an asylum, the new mothers of Vienna once again began dying. In a cruel twist of fate, the institutionalized Semmelweis likewise died of an overwhelming blood infection, most likely contracted through a finger cut he sustained during one of his last medical school autopsies.

By this time, European and American doctors had largely split into two camps: the contagionists, who advocated germ theories, and the sanitarians, or hygienists, who clung to the idea of miasmas—with the added twist that these "poisonous aires" arose from filth and decay.

On the contagionists' side, a number of researchers had glimpsed the presence of microscopic organisms in diseased tissue. But many countered that if, in fact, bacteria existed in the blood and tissues of the sick, they did not cause the disease; rather they sprang— spontaneously—from the dead and dying tissue.[16] Even the renowned Prussian zoologist Christian Ehrenberg, who in 1847 coined the word *bacteria* (from the Greek *baktron*, or "small stick"), maintained that he "view[ed] with disfavor the new-fangled idea that microbes can cause disease."[17]

Still, a few medical researchers entertained suspicions that some types of microscopic life-forms might be behind infectious disease. In 1770 the English physician Benjamin Marten wrote:

The Original and Essential Cause [of tuberculosis], which some content themselves to call a serious disposition of the Juices, others a Salt Acrimony, others a strange Ferment, and others a Malignant Humour, may possibly be some certain Species of Animalculae, or wonderful minute living creatures that, by their peculiar Shape, or disagreeable Parts are inimical to our Nature.[18]

Marten went on to admit that his novel theory would "doubtless seem strange to an abundance of Persons." In fact, such fanciful ideas had already become the subject of popular farce, as witnessed in Samuel Foote's 1768 theatrical comedy *The Devil upon Two Sticks*. In the play, the devil, in the disguise of "Dr. Hellebore," newly arrived president of London Medical College, invites an audience of doctors and dignitaries to look through his microscope for a peep at "little creatures like yellow flies" that he has discovered to be the cause of all illness. Dr. Hellebore's cure?

I administer to every patient the two-and-fiftieth part of a scruple of the ovaria, or eggs, of the spider; these are thrown by the digestive powers into the secretory, there separated from the alimentory, and then precipitated into the circulatory; where finding a proper nidus, or nest, they quit their torpid state, and vivify, and upon vivification, discerning the flies, their natural food, they immediately fall foul of them, extirpate the race of the blood, and restore the patient to health.

As for disposing of the spiders, Dr. Hellebore assures the physicians,

They die, you know, for want of nutrition. Then I send the patient down to Brighthelmstone, and a couple of dips in the saltwater washes the cobwebs entirely out of the blood.[19]

Contagionists remained the subject of ridicule even a century later, when the renowned German pathologist Jacob Henle posed a challenge to his students at the University of Göttingen. "Before microscopic forms can be regarded as the cause of contagion," he declared, "they must be found constantly in the contagious material, they must be isolated from it, and their strength [ability to infect] tested."[20]

In 1876 one of Henle's students, Robert Koch, began fulfilling one after another of his professor's requirements of proof, or "postulates," which today are associated with the younger man's name. Specifically, Koch worked with *Bacillus anthracis*, the anthrax bacterium, and *Mycobacterium tuberculosis*, the bacterium that causes tuberculosis. He repeatedly demonstrated that he could isolate them from the blood or sputum of infected animals and patients, grow them in his laboratory, and finally inject them into healthy laboratory animals to make them sick.

Koch's scientific rival, the French chemist Louis Pasteur, had already shown that airborne microorganisms could interfere with the proper fermentation of wine and beer; he was extrapolating that airborne germs might similarly cause open wounds to "putrefy." Pasteur's ideas so impressed the English surgeon Joseph Lister that in 1865 he began dipping bandages and surgical instruments in a disinfecting solution of carbolic acid—and even pouring the caustic liquid directly into wounds. Though Lister lacked the technical skills of the greatest surgeons of his day, he quickly earned international fame for the unprecedented survival rates of his patients. Over the next decade the soft-spoken and gentlemanly Lister succeeded in convincing a large portion of the Western medical community to begin using antiseptic and aseptic techniques.

Pasteur, meanwhile, realized that before the world could embrace the idea that microbes spread disease by physically moving from person to person, someone had to disprove the enduring belief that microbes arose spontaneously. One of Pasteur's professors, the retired chemist Antoine Balard, helped him design an experiment that did just that. Pasteur partially filled an array of glass flasks with broth; he sterilized both the flasks and the broth by heating them; and he then used a torch to draw the neck of the flasks into a slender curve that allowed in air but prevented the entry of dust particles, with their invisible load of microorganisms. The diagram of "Pasteur's" swan-necked flasks has become one of biology's most familiar.

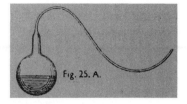

As Pasteur predicted, the broth remained clear until he broke the glass neck, thereby allowing dust-borne microbes to enter. They soon clouded the now-contaminated broth. In this way Pasteur both disproved that microorganisms can arise spontaneously from sterile matter and demonstrated their invisible presence in the air around us.

Pasteur went on to show that heating could kill disease-causing bacteria in food and water. His method of "pasteurizing" milk immediately and dramatically reduced the spread of tuberculosis from infected cows and milkmaids. The germ-theory mindset, together with an emerging understanding of the immune system, also spurred the development of vaccines. Pasteur and others found ways of weakening, or "attenuating," microbes and viruses so that they persisted long enough inside a person to generate immunity but not serious illness. In this way, in 1881 and 1882, Pasteur developed successful vaccines against the anthrax bacterium and the rabies virus. And in 1890 Koch's protégé Emil von Behring developed the first antitoxin vaccines, which primed the immune system to quickly mop up the toxic chemicals produced by *Corynebacterium diphtheriae* and *Clostridium tetani*, the bacteria behind diphtheria and tetanus, respectively.

The early microbe hunters could not discern the profound differences between the ultrasimple "prokaryotic" cell of a bacterium (the subject of this book) and the larger, more sophisticated "eukaryotic" cells of microscopic parasites such as protozoa (the cause of malaria, giardia, and amoebic dysentery) and fungi (the cause of thrush and ringworm), which are much less common.[21] They also mistakenly assumed that viruses were infectious microorganisms that were too tiny for them to see with their most powerful microscopes. We now know them to be nonliving particles of protein-wrapped nucleic acids (DNA or RNA) that enter our cells and there get mistakenly copied ad nauseum, to fuel a new round of infection. But these differences mattered little in terms of preventing infection through sterilization, which not only kills microbes but chemically denatures viruses. Vaccines likewise proved effective against both living organisms and viruses because they work by alerting the immune system to be on guard for a telltale protein or other chemical marker associated with a microbe, virus, or any other potentially dangerous substance, such as diphtheria toxin.

THE SANITARIANS

Ironically, the greatest step in breaking civilization's cycle of deadly epidemics would be made by the camp that didn't believe in germs. The Great Sanitary Awakening of the mid-nineteenth century began when the upper classes of London, Paris, and New York started to worry about the increasingly crowded slums of their cities, where they saw the stagnant cesspools, open sewers, and putrid garbage of the poor as breeding grounds for miasmas. They envisioned poisonous air spontaneously combusting from the filth and decay, then wafting into their genteel neighborhoods by night. Municipal governments began deploying garbage collectors into all corners of their cities and hiring engineers to redesign their sewer systems. Even the clergy joined in the campaign, quoting the eighteenth-century theologian John Wesley, who had urged his middle-class congregants to "teach all the poor whom you visit . . . two things more, which they are generally little acquainted with—industry and cleanliness."[22]

The "cleanliness is next to godliness" school of public health found its most effective champion during the Crimean War, in the renowned "lady with the lamp," Florence Nightingale. In 1854, following scandalous news accounts of the conditions in battlefield hospitals, Nightingale took some thirty women to Scutari (now Üsküdar in modern Turkey) to nurse wounded English soldiers. Though she reported directly to Britain's secretary of war, Nightingale described her work as "cook, housekeeper, scavenger . . . and washerwoman" and her personal mission as banishing the miasmas that were killing far more soldiers than bullets or bayonets.

In a letter to her mother, Nightingale boasts of ridding her wards of cholera by placing a sack of chloridated lime in every corner, as if the mere presence of the disinfectant somehow purged the air of poison. She also attributed declining rates of infection to her practice of "burying dead dogs and white washing infected walls, two prolific causes of fever."[23] Although she misidentified the true enemy in her hospitals, Nightingale's day-and-night scrubbing clearly reduced infection rates, and her 1859 memoir, *Notes on Nursing*, became an international best seller.

But if England gave the world its heroine of hygiene, the United States produced its colonel of cleanliness: Colonel George Waring, a Kaiseresque figure who, to the delight of political cartoonists, waxed his long white mustache thin and straight—with the tips bent up "at attention." A civil engineer by training, Waring rose to the rank of colonel during the Civil War, then spent the postwar reconstruction years as a member of the National Board of Health, traveling the country to advise cities how to build airtight concrete sewer lines that, he claimed, would prevent the escape of miasmic gases.[24] In truth, Waring's sewer systems worked their magic by keeping human waste separate from drinking water. In so doing they provided unprecedented protection from such deadly diarrheal diseases as cholera, dysentery, and typhoid fever. Quite rightly, they became the international model of modern water sanitation.

Waring's contributions to public hygiene didn't stop there. In 1895 he began cleaning up New York City during the brief reign of the reform-minded mayor William L. Strong. Waring found the city's streets "almost universally in a filthy state. In wet weather they were covered with slime, and in dry weather the air was filled with dust. . . . Rubbish of all kinds, garbage, and ashes lay neglected in the streets, and in the hot weather the city stank with the emanations of putrefying organic matter . . . and black rottenness was seen and smelled on every hand."[25]

True to Waring's description, photos of Manhattan in those years reveal streets knee-high in manure, rotting garbage, and dead animals—not only cats and dogs but also the city's workhorses, an estimated fifteen thousand of which dropped in their harnesses each year. Flies looped everywhere. And the few sewer lines in existence dumped untreated waste directly into the harbor and rivers that lapped Manhattan on all sides.

Though he had little patience for politicians, Waring knew how to inspire the public. He adopted the caduceus, or physician's staff, as the sanitation department's symbol, plastering it on hundreds of shiny new garbage-hauling carts. And he dressed his new army of seventy-three hundred street cleaners in gleaming white uniforms that were more evocative of hospital orderlies than of garbagemen. When Waring led his uniformed "White Wings" in their first city parade, the newspapers reported, the crowds came to sneer and stayed to cheer. "His broom saved more lives in the crowded tenements than a squad of doctors," the corruption-fighting journalist Jacob Riis would later write.[26] Waring even enlisted the city's children in his campaign, with a youth sanitation league. Under the campaign slogan "Cleanliness is catching," the New York City Health Department presented badges, certificates, white caps, and prizes to thousands of children who took pledges, sang songs, cleaned up schoolyards, and hectored their families in the ways of "modern sanitation."

With the return of Tammany Hall rule in 1897, Waring resigned. But at the end of his brief reign he could truthfully boast, "New York is now thoroughly clean in every part." The city had become one of the cleanest, if not *the* cleanest, in the world. Most important, in his final report, Waring could assert:

> As compared with the average death-rate of 26.78 of 1882–94, that of 1895 was 23.10, that of 1896 was 21.52, and that of the first half of 1897 was 19.63. If this latter figure is maintained throughout the year, there will have been 15,000 fewer deaths than there would have been had the average rate of the 13 previous years prevailed.[27]

Public water supplies and sewer systems also enabled the widespread adoption of indoor plumbing—both clean water at the tap and a flush toilet in the family bathroom. Already widely adopted in northern Europe, indoor plumbing spread in popularity from the northeastern states to the South and West and from urban regions to rural areas.

The miracle of modern sanitation largely broke the cycle of waterborne epidemics that had begun with the crowding of civilization. Combined with advances in medical hygiene (disinfecting surgical instruments and bandages), it largely accounts for an unprecedented near-doubling in life span throughout the developed world: in the United States, the average life expectancy rose from thirty-eight years in 1850, when the earliest public-hygiene reforms began, to sixty-six years in 1950, just before the widespread adoption of antibiotics.[28]

THE SEARCH FOR MAGIC BULLETS

Koch and Pasteur's proof that specific microbes cause specific diseases spurred a new generation of medical researchers to launch an all-out war on the world of bacteria, with the goal of eradicating them. In doing so they ignored Pasteur's less hardcore view of bacteria: that not all caused harm and that many, if not most, might be beneficial. Pasteur showed, for example, that he could protect lab animals from an injection of deadly anthrax bacterium if, at the same time, he injected them with a ragtag mix of non-disease-causing bacteria from soil and feces. Was that not proof, he asked, that some bacteria actually protect against disease? Pasteur further proposed that the majority of the bacteria found on the skin and in the mouth and intestinal tract of both animals and man were not only benign but essential to life. Pasteur even went so far as to argue that the body's normal complement of bacteria might prove essential to our survival. He encouraged his students to test his idea by attempting to raise lab animals under entirely germ-free conditions, saying, "If I had the time I would undertake this study with the preconceived idea that, in these conditions, life would be impossible."[29]

Pasteur's greatest protégé, the Nobel Prize–winning Elie Metchnikoff, openly scoffed at what he considered his mentor's naïveté. Metchnikoff viewed bacteria as the worst sort of parasites. He blamed the "putrefaction" of bacteria inside the human bowel for senility, atherosclerosis, and an altogether shortened life span. He adamantly advised against eating raw fruits and vegetables ("to prevent the entrance of 'wild' microbes") and predicted that one day surgeons would routinely remove the entire human colon simply to rid us of the "chronic poisoning from its abundant intestinal flora." Till then, he conceded, "it is more reasonable to attack the harmful microbes of the large intestines."[30] At the same time, Metchnikoff's scientifically inclined wife, Olga, took up Pasteur's challenge to test the life-sustaining role of bacteria: her unsuccessful attempt to keep tadpoles alive under sterile conditions no doubt incited some vigorous discussions in the Metchnikoff household.[31] But it was Metchnikoff's view—that the best germ was a dead one—that largely won out in the world of medicine.

The earliest successes in the war against infectious diseases involved vaccines. When vaccines worked, they often worked brilliantly—producing lasting, sometimes lifelong immunity to disease. But like sanitation, vaccines could only prevent disease, not cure it. To treat an active infection, nineteenth-century medicine had little to offer but a few highly toxic antimicrobial drugs, such as mercury and arsenic, which risked killing the patient along with whatever had made him or her sick.

What modern medicine needed, the German pathologist Paul Ehrlich concluded in 1885, was a "magic bullet" that would destroy bacterial cells but leave human cells unscathed. More than an idle daydream, Ehrlich's idea sprang from his own observation that bacterial cells are fundamentally different from our own. The previous year, Danish microbiologist Hans Christian Gram had demonstrated that he could classify all bacteria into two basic groups—now known as gram-positive and gram-negative—based on how their cells absorbed and retained crystal violet dye. The variation, Ehrlich realized, stemmed from differences in the structure of a bacterial cell's semirigid cell wall, a wall not found around the thin membrane that encloses the cells of animals, including humans.

Ehrlich reasoned that if he could find a toxic dye that only bacteria absorbed, it could be the magic bullet of his dreams. He went on to test over nine hundred dyes and related chemicals. In 1908 his compound 606, later called Salvarsan, proved effective against syphilis, a gram-negative bug. But it was far from harmless to patients. Besides causing liver damage, the toxic chemical occasionally forced the amputation of a patient's arm when it accidentally leaked from a vein during injection. Nonetheless, it saved thousands of lives and spurred a number of European chemical companies to sponsor research into finding other chemicals that might be selectively toxic to bacteria. This expanded search produced the sulfa drugs, which like Salvarsan are derived from dyes. Sulfa drugs turned out to be wonderfully effective against a variety of gram-positive bacteria, including *Streptococcus pyogenes*, the cause of strep throat and scarlet fever; and they were moderately effective against *Staphylococcus aureus*, another gram-positive bacterium and a major culprit behind skin and blood infections.

Meanwhile, the world had come close to getting its first true antibiotic—by definition, a microbe-killing compound produced by another microbe. In 1896 the French medical student Ernest Duchesne was studying the competition among microorganisms in his laboratory culture dishes, when he demonstrated that the mold *Penicillium glaucum* easily beat back colonies of the common intestinal bacterium *E. coli*. Further, he showed that if he inoculated lab rats with a combination of the mold and what should have been a lethal dose of typhoid bacteria, the rats survived. Unfortunately, Duchesne did not. He died of tuberculosis before completing his research.

Over the next twenty years, the *Penicillium* mold would catch the attention of several other researchers, most famously the biochemist Alexander Fleming, who in 1928 noticed a mold-contaminated agar plate in which he had been growing colonies of *Staphylococcus aureus*. In one of the most well-known understatements in history, Fleming is said to have remarked to his assistant, "That's funny." For though mold contamination of laboratory petri dishes was common, this mold had killed all the staph bacteria around it.

The next crucial step would be to isolate and purify the mold's bacteria-killing chemical in significant quantities. During World War II the Oxford University chemists Howard Florey and Ernst Chain perfected the process, saving the lives of hundreds of Allied soldiers. Meanwhile, the French-American soil scientist René Dubos opened an even more fruitful avenue of antibiotic discovery. Dubos had long marveled at the profusion and diversity of bacteria found in a clod of dirt and the ability of these microorganisms to degrade, or break down, virtually anything. Indeed, he correctly surmised that we would all be knee-deep in plant and animal carcasses were it not for the recycling abilities of soil bacteria. Soil bacteria, he also reasoned, must use an assortment of biochemical weapons against one another as they jockey for space and nutrients in their crowded subterranean world. And so, in the 1930s he began testing various kinds of harmless soil bacteria for their ability to kill their disease-causing cousins.

In 1939 Dubos isolated the antibiotic gramicidin from the soil bacterium *Bacillus brevis*—this nearly a decade before penicillin came into widespread use. Though too toxic for internal use, gramicidin could be poured into open wounds to prevent or stop infection. Like the partial success of Salvarsan, that of gramicidin spurred a land rush of antibiotic prospecting. Over the next twenty years—the golden age of antibiotic discovery—soil bacteria would supply modern medicine with most of the antibacterial medicines still in use today. In 1962 the Nobel-winning medical microbiologist Macfarlane Burnet opened the new edition of his renowned *Natural History of Infectious Disease* almost apologetically. Noting the revolution wrought by antibiotics, he admitted that "at times one feels that to write about infectious disease is almost to write of something that has passed into history."[32]

PART II

LIFE ON MAN

All the people living in our United Netherlands are not as many as the living animals that I carry in my mouth this very day.

—ANTONI VAN LEEUWENHOEK, 1683

THE BODY AS ECOSYSTEM

The science of ecology, or environmental biology, remained a vague and little-studied concept until the early 1960s, when biologists such as Rachel Carson began describing the looming consequences of environmental disruption. Almost single-handedly, her 1962 bestseller *Silent Spring* made ecology and environmentalism into household words. The same year, Washington University bacteriologist Theodor Rosebury published a treatise on an ecosystem that was more intimate yet far less familiar than Carson's bird-denuded fields and forests. In *Microorganisms Indigenous to Man*, Rosebury summarized all that was then known about the bacterial life-forms populating the human landscape and their interactions with their "environment" for good and bad.[1] In doing so he gave science its first comprehensive census of the human "microflora," along with what little was known about their activities.

Though Rosebury wrote this tome in the driest of academese, the goateed, pipe-puffing professor was infamous on campus for his zest in shaking up classes and seminars with vivid descriptions of the body's microscopic landscape—descriptions that went beyond the merely repulsive to achieve a bizarre kind of beauty. One of his favorite narratives involved a microbe-eye view of the bacchanal depicted in the 1963 Oscar-winning movie *Tom Jones*, a scene frequently described as the sexiest meal ever committed to film:

> Adding my imaginary zoom microscope to [Tony] Richardson's color camera, I was able in my mind's eye to zero in on the little fleshy crevices around Tom's and Jenny's teeth as they ate their meal and to see the turmoil of microbic life there, the spirochetes and vibratos in furious movement, the thicker corkscrew-like spirilla and vibrios gliding back and forth and the more

sluggish or quiet chains and clusters and colonies of bacilli and cocci, massed around or boiling between detached epithelial scales and the fibers and debris of cells and food particles. Like the great and beautiful animals in whose mouths they live, these too are organisms, living things; and I could imagine them, quite like Tom and Jenny, making the most of the sudden accession of nourishment after a long fast.[2]

Rosebury appreciated that the idea of the human body as microbial ecosystem appealed to college students of the 1960s. "Rejecting many of our values, including the whole business of neatness, they see cleanliness as part of the sham of a hypocritical world," he noted. "The myth that germs and dirt are always our enemies is harmful and costly. We ought to get rid of it."[3] The body's tightly knit community of resident "germs," Rosebury argued, was also mankind's strongest bulwark against invasion by their less-well-adapted disease-causing kin. In short, bacteria are part and parcel of a healthy human ecosystem.

Along these lines, Rosebury described human birth as the emergence of a prime chunk of microbial real estate, its colonization involving the merger of two vastly different kingdoms of life. If all goes well during the first few hours, days, months, and years of life, the end result will be the formation of a protective microbial force. This "national guard" will fan out across the child's skin and mucous membranes, from her waxy scalp through the tunnels of the respiratory and digestive tracts and down to the sweaty crevices between her toes. And odds are, Rosebury stressed, things will go well. This baby will thrive in a world awash with microbial life, her instincts compelling her to finger and mouth every bacteria-smeared surface within reach. A similarly instinctual compunction, he suggested, spurs parents, siblings, and family pets to add their germy kisses, cuddles, and licks.

Preparations for the human-microbe alliance begin before birth. Midway through pregnancy, a hormonal shift directs the cells lining a woman's vagina to begin stockpiling sugary glycogen, the favorite food of the plump, sausage-shaped bacteria called lactobacilli. By fermenting the sugar into lactic acid, these bacteria lower the pH of the vagina to levels that discourage the growth of potentially dangerous invaders. These threats include the occasional intestinal bacterium that might stray from anus to vagina, then overgrow and spread into the uterus to cause a pregnancy-threatening infection. The acid-secreting lactobacilli also provide partial protection against the sexually transmitted bacteria *Neisseria gonorrhoeae* and *Chlamydia trachomatis*, which can cause blindness in newborns infected during passage through the birth canal.[4]

Some vaginal lactobacilli also produce hydrogen peroxide, essentially the same fizzy sanitizer that mothers pour on scraped knees.[5] These extra-aggressive lactobacilli are particularly effective at beating back the growth of *Streptococcus agalactiae*, or group B strep. Commonly found in the vaginas of women who lack hydrogen-peroxide-producing lactobacilli, group B strep remains a leading cause of infant mortality.[6] Each year in the United States it causes thousands of life-threatening cases of pneumonia, meningitis, and blood infections, primarily in babies under one month of age. Because the natural protection of lactobacillus is far from foolproof, Western obstetricians routinely put women who test positive for group B strep on antibiotics during labor. Ironically, previous courses of antibiotics (for unrelated conditions) are often to blame for allowing group B strep to move into the birth canal in the first place—for these drugs tend to disrupt the woman's protective lactobacilli.

INTO THE MOUTHS OF BABES

Over the course of his life, the eclectic Rosebury researched everything from bacteria-generated "fecal body odor" (the smell of farts) to the threat of bacteria as biological weapons. Still, his faculty position within Washington University's School of Dentistry kept Rosebury's primary focus on the microbial ecosystem of the human mouth. His deep knowledge of this niche gave us our first comprehensive description of the colonization and maturation of the microbial community found behind the lips of a healthy human.

Typically, the infant mouth's first inoculation of bacteria includes a generous sampling of the lactobacilli present in the mother's birth canal. With the first gulp of breast milk, these lactobacilli are joined by millions of bifidobacteria, a related group of acid-producing microbes.

These stubby, forked microbes mysteriously appear in and around the nipples of a woman's swelling breasts during the eighth month of pregnancy. There the bifidobacteria secrete a potent combination of acids and antibiotic chemicals to repel potentially dangerous microbes such as *Staphylococcus aureus* (with its nasty tendency to infect wounds as small as a baby pimple). Their sudden appearance on the breasts during late pregnancy mystified Rosebury, for these bacteria are anaerobic— that is, "oxygen shunning," or unable to survive in open air. It turned out that they grew deep in the oxygen-free chambers of the milk ducts and leaked out with the first invisible drops of colostrum, or "pre-milk." Though the bifidobacteria themselves perish in the open air, they leave behind acids that linger for hours on the breast and in a baby's mouth. They join with the lactobacilli in helping select the mouth's first permanent settlers. These include the acid-tolerant *Streptococcus salivarius*, beadlike bacteria that appear on a baby's tongue during the first day of life. Using strong adhesins, or biochemical grappling hooks, this "good" strep anchors its chainlike colonies on the tongue's roughened surface and there remains one of the predominant species of a healthy mouth. Several other streptococci, such as *Strep. oralis* and *Strep. mitis*, settle in over the first week of life. So too do one or more kinds of neisseria bacteria, which appear as paired orbs sprouting across the gums, palate, and inner cheeks.[7] If the newborn is fortunate, her neisseria population will include the fuzzy brown globes of *Neisseria lactamica*, which thrives on lactose, or milk sugar. Early colonization by this species builds strong immunity against its felonious cousin, *Neisseria meningitidis*, the most common cause of bacterial meningitis, a potentially deadly inflammation of the membranes covering the brain and spinal cord.[8]

Where do all these early colonizers come from? By identifying the specific subtypes of bacteria in children's mouths, researchers have found that the vast majority trace directly to the mouths of their mothers—always within easy reach of a nursing baby's tiny fingers.[9] The maternal antibodies still circulating in a newborn's blood (passed during pregnancy) may further encourage the growth of microbes that they recognize as "their own." Older siblings, especially older brothers, rank second in contributing to the baby's bacterial colonization, perhaps because of their less-than-perfect personal hygiene.

As the baby settles into a pattern of nursing (or drinking formula), the bacterial population inside her mouth noticeably ebbs and flows with the milky tide. With each meal the resident microbes burgeon in number, then subside again. The mouth's salivary glands secrete a more constant but less intense source of bacterial food—a watery mix of proteins, sugars, and minerals. The epithelial cells that line the inner cheeks and tonsils dole out mucins, a slippery kind of glycoprotein, or "sugar" protein. Mucins nourish resident bacteria but keep them from directly attaching to and damaging the delicate epithelium.

The mouth's first wave of aerobic bacteria consumes enough oxygen to create an underlying zone where anaerobic bacteria can thrive. By the time a baby is two months of age, a microscopic close-up of her gums will reveal clusters and chains of various veillonella bacteria and long branching cells of several kinds of actinomyces. These anaerobic bacteria feed on the biochemical waste products of their microbial neighbors and so help stabilize their ecosystem as it grows crowded.

Until the oral ecosystem reaches maturity, the baby remains vulnerable to thrush, an overgrowth of the fungus *Candida albicans*. Appearing as creamy raised patches, colonies of candida send out hyphae, or filaments, that penetrate and inflame tissues to produce open sores. As the bacteria of the mouth grow in density and diversity, the candida population generally dwindles to a scattering of isolated cells. These can reblossom into invasive colonies should something disrupt the mouth's ecological balance. Often that something is a course of antibiotics given to the baby or the nursing mother.

Another wave of immigrants arrives with the sprouting of a baby's first teeth. The first of these is *Streptococcus sanguis*, a bacterium unsurpassed in its ability to cling to the smooth enamel face of small incisors. The eruption of the first molars brings *Streptococcus mutans*, infamous for causing cavities. Fortunately, *Strep. mutans* tends to be more easily dislodged, both by the bacteria-loosening antibodies in saliva and by sheer force when a person chews solid food or brushes with a toothbrush. It most often causes problems when it nestles into a molar's deep fissures or into the tight spaces that form when teeth crowd together. Still other kinds of tooth bacteria tend to piggyback on top of the enamel-clinging streptococci. Together they form layered communities, or biofilms, that reassemble themselves within hours of a good brushing. By middle childhood, the diversity inside a healthy mouth surpasses a hundred species, out of a possible five to seven hundred, and their total number tops 10 billion bacteria (twice the world's human population).

Rosebury also studied the bacterial ecosystem of the nasal cavities, which are connected to the mouth via the upper respiratory tract. With a baby's first sniff, she draws in thousands of bacteria-laden dust particles. These inhaled bacteria immediately encounter a toxic brew of biochemical weapons. Each droplet of nasal mucus is chock-full of lysozymes and defensins, molecules that punch holes in a bacterial cell wall. The gluey mucus of the nose also forms a tar trap that clumps incoming dust particles for delivery to a field of sweeping cilia. The cilia, which resemble microscopic strands of hair, are live extensions of the epithelial cells that line the nasal cavity. Their barbed shafts beat ceaselessly, pushing through the syrupy nasal mucus and propelling trapped microbes to the back of the throat, where they can be either swallowed or coughed out through the mouth.

Despite these defenses, the baby's nose doesn't remain sterile for long. Within two days of a baby's birth, a select group of tenacious bacteria have settled in. *Staphylococcus aureus*, or "golden staph," often predominates in the first month of life, but by six months it tends to be crowded out by either *Streptococcus pneumoniae* or *Haemophilus influenzae*. *Moraxella catarrhalis* arrives by the end of the year. Though harmless in the nose, any one of these bacteria can cause an ear, sinus, or lung infection when it strays and overgrows in the wrong place. Subtle differences in the microbe-killing chemicals in a baby's mucus may determine which of these organisms take up permanent residence. Few people end up harboring all four.

In the 1960s bacteriologists discovered a more unusual troublemaker in the noses of some infants. The impossibly small, squishy cells of *Mycoplasma pneumoniae* rank among the smallest living organisms. They belong to the bizarre family of bacteria known as mollicutes. The largest mollicutes barely reach one-fifth the size of a more typical bacterium such as *E. coli*. But their greater distinction is their lack of a cell wall. This seeming vulnerability renders mollicutes impervious to the many antibiotics and immune system chemicals that target a bacterium's cell wall. Today pediatricians know that *M. pneumoniae* mysteriously ebbs and flows in the general community in five-year cycles, causing wintertime epidemics of painful ear infections and so-called "walking pneumonia," both of which usually run their course without antibiotic treatment.

LIFE ON THE SURFACE

In 1965 New Zealand microbiologist Mary Marples published a companion volume to Rosebury's *Microorganisms Indigenous to Man*. Unprecedented in its detail, her *Ecology of the Human Skin* explored how "climatic" factors determine what grows where on a person's skin. By *climate* she meant not only such influences as humidity and heat but also the kind of clothing people wear, their personal hygiene, and even genetically determined traits such as how much they sweat. Marples viewed the human body as not one landscape but many—from the veritable desert of the arms and legs to the temperate woods of the scalp and the lush and humid jungles of armpit and groin.[10]

As with the mouth and nose, the colonization of the human skin begins during birth, with the lactobacilli in the mother's birth canal. These protective bacteria contribute their lactic acid and hydrogen peroxide to the bacteria-killing enzymes in the creamy vernix that covers the emerging baby. Like the bifidobacteria of breast milk, lactobacilli don't survive long in open air. But the mantle of acid they leave behind helps select their successors from the assortment of bacteria that rain down on the newborn aboard dust particles and airborne flecks of skin shed by those around her. The hands and exhalations of parents and birth attendants likewise transfer bacteria to the baby's skin.

The winners in this land rush for territory almost always include plump, cream-colored spheres of the acid-loving bacterium *Staphylococcus epidermidis* (*Staph. aureus*'s better-behaved cousin). Two hours after birth, by which time the mother's vaginal lactobacilli have vanished, the baby's starter set of *Staph. epidermidis* have multiplied exponentially. If you could examine her skin with an electron microscope, you would spot three to sixteen *Staph. epidermidis* on every skin cell, looking like a scattering of fuzzy tennis balls on a rough cement court.

By the end of the first day of a newborn's life, *Staph. epidermidis* is joined by a half dozen or so kinds of coryneform, or "club-shaped," bacteria. The most oxygen-dependent of this group—*Corynebacterium jeikeium* and *C. urealyticum*—fan out across the skin's surface. The more adaptable *C. amycolatum*, *C. minutissimum*, and *C. striatum* settle into the relatively airless depths of hair follicles, and the oil-loving *C. lipophilicus* nestles into the waxy sebum that covers the baby's scalp. Both *Staph. epidermidis* and the corynebacteria tolerate salt levels high enough to pickle most microbes and so will thrive in the presence of sweat. Twenty-four hours after birth, the suburbanization of the baby's skin has gone far, with more than a thousand bacteria per square centimeter (six thousand per square inch). This rapid pace of growth continues through the second day, surpassing ten thousand per square centimeter at forty-eight hours and hitting the hundred thousand mark by six weeks. At such densities, the skin's largely aerobic pioneers begin to deplete the limited oxygen inside hair follicles, glands, and other crevices, preparing them for a second wave of settlers: branching colonies of anaerobic propionibacteria. For the most part, the skin's microflora stabilize by middle childhood. Though family and friends continually trade microbes, fewer and fewer of these later visitors remain for long. The skin's tight-knit climax community of bacteria has effectively filled all niches, leaving little room for newcomers, harmless or otherwise.

"Although pathogenic organisms constantly alight on the skin, they find it a most unfavorable environment," Marples observed in 1969. "The 'self-sterilizing' capacity of the skin does not, as the term suggests, seem to be an attribute of the skin itself. Rather, it is the characteristic displayed by all well-developed ecosystems." Among the densest communities Marples studied were the prairies of corynebacteria that grow across the humidity-prone regions of groin, neck, and toes. The more corynebacteria, the fewer candida, microsporum, trichophyton, and malassezia—the culprits behind such fungal skin infections as diaper rash, athlete's foot, ringworm, and seborrheic dermatitis.

With adolescence, the pimple-inducing *Propionibacterium acnes* lives up to its name: it becomes trapped and overgrows inside overactive oil glands. Around the same time, the corynebacteria of the armpit begin feasting on an entirely new source of food: secretions of steroidal hormones, primarily androstenone in men and androstenol in women. The products of their digestion produce the unmistakable reek of budding adulthood. Though consumers spend millions each year on products that squelch this microbial action, our armpit bacteria may be performing the role played by scent glands in other animals: producing pheromones, or sex attractants, from the otherwise odorless steroids.[11] Unquestionably, the bacteria-generated odor from the armpits of a teenage boy can overpower that of a young woman (or anyone else). But hers has the intriguing pattern of ebbing and flowing over the course of her monthly menstrual cycle, her scent peaking just before she ovulates—that is, when she is at her most fertile.

The high estrogen levels of female adolescence also foster the growth of vaginal lactobacilli. Though they are less dense now than during pregnancy, their levels are normally high enough to discourage the growth of the intestinal bacteria and fungi that get introduced into the vagina during intercourse. A girl does not have to be sexually active to accidentally introduce a small number of these troublesome microbes into her vagina—she may do so, for example, when she wipes after a bowel movement. But they seldom cause problems as long as her lactobacilli population remains in place. Because antibiotics, especially broad-spectrum antibiotics, tend to disrupt this ecological balance, they frequently trigger either "yeast" infections (caused by the fungus *Candida albicans*), bacterial vaginosis (caused by intestinal bacteria), or a maddening cycle of one followed by the other.

Like Rosebury's, Marples's intimate explorations of the human body proved weirdly fascinating to the general public. A 1969 feature story on her work in *Scientific American* magazine[12] prompted the English poet W. H. Auden to write a paean to his own body microflora, which he published as "A New Year Greeting." It reads, in part:

> *A Very Happy New Year*
> *to all for whom my ectoderm*
> *is as Middle-Earth to me.*
>
> *For creatures your size I offer*
> *a free choice of habitat,*
> *so settle yourselves in the zone*
> *that suits you best, in the pools*
> *of my pores or the tropical*

forests of arm-pit and crotch,
in the deserts of my fore-arms,
or the cool woods of my scalp

Build colonies: I will supply
adequate warmth and moisture,
the sebum and lipids you need,
on condition you never
do me annoy with your presence,
but behave as good guests should,
not rioting into acne
or athlete's-foot or a boil.[13]

LIFE ON THE INSIDE

If you could ball up the skin's 100 billion or so resident bacteria, they would fit inside a medium-size pea. By contrast, Rosebury estimated that the 15 trillion–odd bacteria cells lining an empty digestive tract would fill a ten-ounce soup can to overflowing. To this total, he added upward of 100 trillion bacteria massed and ready to evacuate inside a typical bowel movement.

But for all their mind-boggling numbers, the intestinal microflora had received little attention before Rosebury published his 1962 census. Most famously, in 1885 the German pediatrician Theodor Escherich isolated "Bacterium coli" from the stool of newborns. Renamed in Escherich's honor, *Escherichia coli* remains the best known of all the intestinal bacteria, and in Rosebury's day it was still thought to be the colon's predominant species. It was the one bacterium that consistently showed up in stool cultures and sewage-contaminated water supplies. In truth, *E. coli* was simply the easiest of the intestinal bacteria to grow outside the body.

Microbiologists had long suspected that the intestines contained a huge host of uncultivated and unidentified organisms. For when they examined diluted stool samples under the microscope, they realized that they were counting far more bacterial cells—the equivalent of 100 billion per gram—than ever grew out into colonies when deliberately cultured on petri dishes. As it turned out, the touch of oxygen proved deadly to the vast majority of the intestinal bacteria. They were "strict" anaerobes—in contrast to the switch-hitter *E. coli*, a "facultative" anaerobe that thrived with oxygen or without. It also seemed likely that the nutritional requirements of most intestinal bacteria differed from the usual mixture of proteins and sugars that constituted the menu in standard lab growth media.

As tempting as it was to ignore this persnickety and little-understood population, a small handful of microbiologists began to appreciate its profound importance, thanks to information gleaned from gnotobiology, the study of lab animals birthed and raised under germ-free conditions. Without bacteria, the intestinal tracts of these animals remain underdeveloped, the lining unusually thin and injury-prone in places and grossly distended in others. Just keeping them alive to adulthood requires bolstering their diets with vitamins, essential amino acids, and extra calories to replace those normally supplied by their intestinal bacteria. The animals also prove unusually vulnerable to naturally occurring toxins in their food, perhaps because they lack one or more kinds of bacteria necessary for breaking these poisons down into harmless substances. Most dramatic of all, a germ-free animal's immune system remains in a kind of dormant state, leaving it unusually susceptible to deadly infections should a stray bug slip into its sterile environment.

The puzzle was intriguing, but funding for isolating and studying the dimly glimpsed members of the colon's exotic ecosystem proved scarce to nonexistent. Government health agencies simply didn't fund the study of *harmless* bacteria. They lavished their grants on those studying disease-causing invaders such as salmonella, shigella, and other major causes of so-called food poisoning. Then NASA came calling.

BUGS IN SPACE

The space agency's sudden interest in the body's microbiota in general and in anaerobic intestinal bacteria in particular began with a quirky report presented to an audience of NASA test pilots and medical staff at the end of April 1964. The NASA flight surgeon Charles Berry must have thought he had enough to worry about, what with predictions that eyeballs would explode in zero gravity (thankfully disproved) and that bones and muscles would turn to mush after prolonged periods of weightlessness. Now here was a scientist proposing that the biggest danger to returning astronauts might be the kisses delivered by their wives when the men emerged from their prolonged isolation from Earth's germy atmosphere. "Microbic shock," Don Luckey had called it during his presentation to NASA's Nutrition in Space conference at the University of South Florida.[14] "Luckey's Fatal Kiss" was what appeared in newspaper headlines the next day.

Luckey, a gnotobiology pioneer, already knew what happened when you isolated conventionally raised rats into small groups inside hermetically sealed chambers and then fed them sterilized food and water—a situation not unlike that of astronauts living on freeze-dried food and Tang during spaceflight. After two months, the normal diversity of bacteria in the animals' intestines dwindled from upward of a hundred different species down to just one or two.

"Clearly, our normal microflora is not so much indigenous as a continual flow of new immigrants," Luckey had explained. Without that influx a richly diverse ecosystem deteriorated toward monoculture. Depending on what won out, this loss of diversity might itself prove deadly. Luckey raised the example of E. coli. When tempered by the presence of some other kind of intestinal bacterium, he said, E. coli remained harmless. But by itself it proved deadly.[15] Even if some utterly harmless bacterium won out, the result might be a "lazy" immune system. In his own experiments Luckey had observed how easily germ-depleted animals became sick and died when reintroduced to a normal colony of rats.

That was where Luckey's "fatal kiss" came in. A lunar mission would last approximately two weeks. Add to that a monthlong postflight quarantine (to ensure that the astronauts hadn't picked up some dangerous sort of moon bug). The men would then emerge from their isolation—bacterially depleted and immune compromised. Their wives would rush into their arms and lock lips. "There is little question that one or more types of microbic shock will be a problem to future astronauts," Luckey concluded. "Some may be so subtle that they would be of only academic interest. Others may lead to disease and death."

Luckey's predictions transformed the "merely interesting" micro-flora of the human body into a life-or-death issue. NASA's Berry quickly found the funds for Luckey to study the microflora of primates that had been maintained on a yearlong diet of dehydrated and irradiated space food. He also piggybacked an exhaustive microbial census on a previously planned study of the physical and psychological effects of thirty-four days of confinement under near-space conditions on six test pilots. The census included throat, mouth, and skin swabs taken ten times over the course of the isolation, along with the men's daily bowel movements; all of these specimens were passed through a double-doored tube that separated the test pilots from microbiologists Lorraine Gall and Phyllis Riely. Over the course of the study the two women used more than 150,000 petri plates and test tubes and more than 10,000 microscopic slides. But their studies were limited to known microbes—that is, those amenable to laboratory culture, which included a number of the less fussy anaerobes.[16]

As expected, they found that the total number of bacteria on the astronauts' skin increased with their confinement and limited opportunity to wash; several potentially troublesome kinds of staph and strep bacteria rose to predominance. None of these changes resulted in illness. However, a marked shift in the astronauts' intestinal flora produced a more immediate problem in the confined airspace of the test chamber: an outbreak of flatulence noxious enough to trigger an urgent directive to NASA nutritionists to explore the influence of diet on gas-producing intestinal bacteria. Still, all six astronauts emerged from their experimental chamber healthy, and they remained so over the following month. The study left unanswered what greater changes, if any, might occur among fewer astronauts living in closer confinement for longer periods of time.

In 1966 NASA promoted Berry from "top doc to the astronauts" to the agency's chief of biomedical research. In addition to safeguarding his men from microbial shock, he also faced the task of ensuring that their native bacteria didn't confuse the planned search for lunar life. For NASA scientists would not be able to distinguish moon bugs—if they existed—from Earth microbes unless they had a complete accounting of every organism "contaminating" the astronauts, their space suits and equipment, and everything they touched. Berry led the first systematic cataloging with a tally of skin and mouth microflora taken

before and after the previous two Gemini missions. He hired microbiologist Gerald Taylor to lead a more comprehensive cataloging of
crew microflora for all the Apollo missions.

As for dangerous changes in the astronauts' microflora, Taylor found
that the early Apollo missions agreed with *Candida*, a troublesome
yeast that turned up in abundance in the mouths and stool of many returning Apollo astronauts. So other than an easily treated case of
thrush, he predicted that nothing too worrisome would result from the
longer isolation involved in the upcoming Apollo 11 moon mission.
When the astronauts Buzz Aldrin, Neil Armstrong, and Michael Collins
emerged from their three-week postlunar quarantine in August 1969,
no one stopped their wives from kissing them, though Berry made sure
they avoided the usual crush of reporters and photographers by releasing the three men to their families in the dead of night.

Microbial shock remained on the minds of NASA microbiologists
and flight surgeons, for Skylab missions of several months' duration
were already in the planning. NASA's budding détente with the USSR
space program exacerbated their fears, as the Soviets were reporting far
more dramatic and potentially problematic changes in their cosmo-
nauts' microflora than any that had shown up in NASA studies. Most
mystifying of all, the Soviets had noted a virtual takeover of the intestinal tract by a small handful of drug-resistant and toxin-producing
strains of bacteria.[17]

Berry lobbied hard for money to conduct a full-dress, fifty-six-day
simulated Skylab mission inside a high-altitude test chamber at Johnson Space Center. But once the moon race was won, Congress slashed
NASA's generous annual budget by hundreds of millions of dollars.
Berry got just enough money for Taylor to conduct a superficial
overview of crew microbiota, with a little left over to farm out a more
in-depth exploration of the men's intestinal bacteria.[18] Still, NASA's
leftovers would be enough to jump-start an unprecedented exploration
of the anaerobic dark matter of the human microcosm.

WHERE NO BIOLOGIST HAS GONE BEFORE

Peg Holdeman recalls the excited look on Ed Moore's face one day in late 1971 when he handed her NASA's eye-popping request for research proposals. A hundred thousand dollars. "To us it was a humongous amount of money and perhaps the answer to our pie-in-the-sky dreams," says Holdeman. For years Holdeman and Moore had been scavenging for funds to study the finicky anaerobic bacteria that dominated life on man—*on* being the right word to describe the outside-in tube of the human digestive tract. They had met six years before, at a conference of the American Society for Microbiology. At the time, Holdeman was in charge of culturing and identifying anaerobic bacteria that state health departments sent to the U.S. Communicable Disease Center (predecessor to the Centers for Disease Control). Though few medical professionals of that day believed that anaerobic bacteria could cause disease, some wanted to know the identity of the mysterious microbes they glimpsed in the blood and tissues of their patients— microbes they could not themselves culture. Holdeman loved the challenge but felt thwarted by the inadequacy of her equipment and staff. At best, her lab could identify only a small fraction of the anaerobic bacteria sent their way.

Meanwhile, Ed Moore, a professor at the Virginia Polytechnic Institute, had ample funds for studying anaerobic microbes—just not those living inside humans. Both the U.S. Department of Agriculture and the cattle industry had been funding anaerobic research since the discovery, in the 1930s, that anaerobic bacteria powered the production of beef and milk. Specifically, the anaerobic bacteria of a cow's rumen, a kind of "antechamber" to the stomach, broke down the otherwise indigestible plant fibers in its feed, supplying the majority of the animal's calories as well as an abundance of vitamins and other nutrients. By understanding the dynamics of this process, researchers such as Moore hoped to ramp up its efficiency—causing cattle to produce more milk and meat for less hay and grain. He had already improved on a remarkably efficient technique for culturing anaerobic bacteria. It involved growing them on a specially enriched growth medium inside a sealed tube from which all oxygen had been purged.[19]

Moore's interest in human intestinal bacteria had begun as a wholly unfunded side project: a graduate student had inoculated some of Moore's anaerobic tubes with human stool instead of the usual rumen fluid. Under Moore's direction, the student had succeeded in culturing around 80 percent of the bacteria glimpsed in the stool samples, a record amount. But Moore had no funding to sponsor further study. So the wealth of newly cultured species remained undescribed and un-named.[20]

Moore related all this to Holdeman at the 1965 microbiology conference. The two microbiologists suspected that 99 percent of intestinal bacteria would turn out to be anaerobic and that this bacterial community had a profound effect on human health, both good and bad—the latter when some of its members ended up in the wrong place, as when they contaminated wounds and surgical incisions. "It was like two wayward souls getting together," recalls Virginia Tech microbiologist Robert Smibert. "[We] just sat there and listened to them talk. Before we knew it, they had come up with what they wanted to do with anaerobes. So Peg left CDC and came up to the vet science department."[21] That, says Smibert, was the beginning of what would become Virginia Tech's world-renowned Anaerobe Lab.

Still, getting money to study the little-understood world of the body's anaerobic bacteria proved difficult. The first big break came with the astronaut intestinal flora grant at the end of 1971. Within a month they won a second, larger grant from the National Cancer Institute. Recent studies had suggested that abnormal intestinal flora might promote colon cancer, enabling Holdeman and Moore to convince the agency that science needed to define "normal." The simultaneous start of two major studies produced a madhouse of activity for Virginia Tech's newly assembled Anaerobe Lab. When Moore wasn't flying to Houston to pick up fresh astronaut stool, he and Holdeman were driving up to Baltimore to camp out in a cheap motel and wait for a call from the medical examiner: a large part of their Cancer Institute study involved confirming that intestinal microbes isolated the easy way— from stool samples—were the same as those that lived in direct contact with the human colon. Asking a surgeon to take intestinal scrapings during abdominal surgery wasn't an option, because preoperative antibiotics kill off intestinal bacteria by the billions (and raze some species more than others). Their solution was to get scrapings from freshly dead but otherwise healthy victims of Baltimore's mean streets — primarily homicide victims and traffic fatalities. As it turned out, all the microbes growing in contact with the intestinal lining could be found in human stool.

Around the same time, Holdeman and Moore's work for NASA helped the space agency put fears of microbial shock to rest, at least for near-Earth space missions spanning months rather than years. Their studies of astronaut microflora produced one intriguing result. Over the first three weeks of the Skylab simulation, all three astronauts experienced a dramatic spike in their intestinal populations of the hydrogen-gas-producing anaerobe *Bacteroides thetaiotaomicron,* or *B. theta.* The bug shot up in number from a normal count of around 2 trillion, or 2 percent of the intestinal flora, to as high as 26 trillion, or more than 25 percent. By the six-week mark *B. theta* levels had dropped back to normal for astronauts Robert Crippen and Karol Bobko, but they remained sky-high for crew member William Thornton. Had anything unusual happened over the first six weeks? Moore asked. The crew medical officer laughed in response to the question. The three astronauts had staged a near-rebellion in the first three weeks, so frustrated were they about a string of equipment snafus and a perceived lack of cooperation from "ground control." Just after the three-week mark, a teleconference was held to address the issues; Bobko and Crippen came away satisfied, but Thornton remained in a rage, owing to an unresolved dispute over the size of the meals he had requested. Could his emotional upheaval be causing his *B. theta* spike? Moore and Holdeman wondered. "We would have dismissed the idea as far-fetched if not for what we encountered later," recalls Holdeman.[22] She and Moore observed similarly dramatic spikes in *B. theta* on two other occasions: once during a diet study, when a volunteer became embroiled in a workplace spat that almost got her fired; and a second time in the intestinal flora of a nineteen-year-old Baltimore woman, a murder victim whose husband had shot her after a prolonged beating and chase.

Intrigued, Holdeman and Moore looked for similar shifts in graduate students under stress (taking stool samples on the day of their oral examinations), but they found no dramatic changes. "Perhaps you had to be really, really, really angry or scared," Holdeman muses. "But our study review committee would hardly allow us to go chasing after graduate students with knives." In any case, the B. *theta* spikes observed in enraged astronauts and warring office workers produced no apparent harm.

Both NASA and the Cancer Institute found their work useful. More important for their long-term goals, the twin studies enabled Holdeman and Moore to isolate 150 to 200 new kinds of human intestinal bacteria. They identified all of these bacteria down to the genus level, then concentrated their time and money on a complete physical and biochemical description of the dozen most common species. These newly named bacteria included an entirely new genus, *Coprococcus*, comprised of bacteria that broke down toxic plant chemicals called phloroglucinols, and a variety of *Rumenococcus*, a group previously unknown in humans but familiar for helping cattle digest their hay.[23]

By the time Holdeman and Moore married in 1985, their working partnership had given science a comprehensive overview of the microbial community of the human colon, or large intestine. By their optimistic estimate, they could account for more than 90 percent of its members and all of its predominant species. And though the individual roles and interactions of these bacteria remained mysterious, it had become clear that, in aggregate, this community functioned much like that inside a cow's rumen. In essence, the bacteria-packed human colon was a remarkably efficient biofermentation center, where resident microbes extracted calories and nutrients from otherwise undigestible plant food—all the while sharing a significant portion of this bounty with their host.

THE INNER TUBE OF LIFE

While two microbiologists gave science a near-complete census of our intestinal bacteria, an understanding of how these microbes colonize their environment—and even adapt this environment to their needs—continues to emerge gradually. Like the colonization of the mouth and the skin, that of the human digestive tract—home to 99 percent of the body's microflora—begins during birth, starting with the lactobacilli encountered in the birth canal. As the baby's head crowns, it compresses the mother's rectum, pushing out a small amount of stool. Though doctors and nurses move quickly to wipe away the offense, their squeamishness may run counter to nature's purpose — an immediate and direct inoculation of the newborn with the mother's own intestinal bacteria. If so, it's no coincidence but rather the result of natural selection that a newborn's head typically faces in the direction of its mother's rectum when its head first emerges and remains there until the next contraction delivers the shoulders and the rest of the body.[24] This head-to-anus juxtaposition ensures that, of all the billions of microbes the baby will meet in its first day of life, the first will be those to which its mother's immune system has already developed protective antibodies. (A temporary supply of these antibodies have already passed to the fetus through the placenta.) A chaser of breast milk delivers the second wave: millions of bifidobacteria.

All incoming microbes, before they reach the intestines, must pass through the antechamber of the baby's stomach. In the stomachs of older children and adults, high levels of hydrochloric acid present a microbe-killing barrier. But acid secretion doesn't begin in earnest until around three months of age, building gradually to adult levels over several years. This delay leaves open a welcoming door for colonization of the stomach and intestinal tract during early life. Say, for example, a baby meets the stomach bug *Helicobacter pylori* during this period. Children typically do so on the hands or lips of someone already colonized. Once swallowed, this whirligig of a bacterium drills into the mucus layer that will protect the stomach from the hydrochloric acid bath to come. As the acid-producing cells of the stomach mature, some strains of *H. pylori* inject proteins that direct the cells to lower the acidity to levels that are more tolerable to *H. pylori*, but still caustic enough to kill most other kinds of microbes. In this way *H. pylori* has maintained a virtual monopoly over the human stomach for some sixty thousand years. The strain of *H. pylori* carried by a particular family or population can even be used to trace their ancestry as far back as the ancient migration patterns that began when *Homo sapiens* first hiked out of Africa.[25]

On the positive side, this lowering of stomach acid will protect against acid reflux and esophageal cancer in adult life. On the downside, in a small minority of those infected, *H. pylori* triggers inflammation serious enough to produce gastric ulcers and stomach cancer in later life. It's an enduring paradox of modern medicine that *H. pylori* did not begin producing ulcers until the early to middle nineteenth century, just when it was beginning to disappear from human stomachs due to improved water sanitation and the use of early antibiotics such as bismuth water. Specifically, in the 1830s previously healthy young European women began dying from ulcers so severe that they perforated their stomachs.[26] The suddenness and agonizing nature of these women's deaths made clear that this was indeed a new rather than a previously overlooked disease. As the affliction spread from women to men and from Europe to North America, it remained primarily a disease of the urban classes—leading to the conclusion that ulcers were caused by the stress of modern life. At the dawn of the twentieth century medical experts considered farmers and other rural folk largely immune thanks to "the outdoor life and comparative freedom from worry . . . that enable them to digest articles that would ruin the stomach of a bookkeeper or his employer."[27]

Not until the 1980s did the Australian pathologists Robin Warren and Barry Marshall finally convince a skeptical medical world that the whip-tailed bacterium they had discovered in ulcer biopsies was somehow linked to the damage.[28] Yet today gastric ulcers remain virtually unknown in undeveloped regions of the world such as Africa, where most people become colonized in infancy. It may be that delaying or disrupting *H. pylori* colonization with water sanitation or antibiotics has somehow altered the immunological "truce" that this microbe forged with our immune system over thousands, possibly millions, of years.

In any case, *H. pylori* is rapidly becoming extinct in the Western world—possibly leaving the stomach open for colonization by some other, less coevolved bug. Today fewer than 10 percent of children in North America and western Europe carry *H. pylori*, though it can be found in around 30 percent of their parents and the majority of their grandparents.[29] The good news is that this decline has brought a dramatic decrease in both gastric ulcers and stomach cancer. The bad news is that its disappearance may underlie the unprecedented increase in acid reflux disease and esophageal cancer that's taken place over the last thirty years; esophageal cancer is now one of the most rapidly rising causes of death in the developed world.[30]

Proceeding past the antechamber of the stomach, surviving microbes enter the switchback labyrinth of the small intestine, where a forest of fingerlike extensions, called villi, maximize the surface area through which nutrients can enter the bloodstream. The low acidity, or neutral pH, of the small intestine proves ideal for bacterial growth. But few microbes linger for long, as powerful contractions produce a torrent of liquefied food that dislodges all but the most tenacious.

The small intestine is where incoming microbes most directly engage the infant's dormant immune system. In places, the undulating villi part to reveal barren hillsides, their surfaces a mosaic of oddly flattened and pitted cells. Called Peyer's patches (after the seventeenth-century Swiss anatomist Hans Conrad Peyer, who first described them), these domelike structures overlie the immune system's most important training academies. The pits on the surface of these flattened cells are pockets that continually snag passing bacteria (both live and dead). Like revolving doors, these pockets migrate to the cells' inner surface, ushering their microbial passengers into the lymph tissue below.[31]

The internal structure of the Peyer's patch resembles that of the lymph nodes that will mature in the infant's neck, groin, and armpit. But if lymph nodes resemble war rooms where immune cells learn what to attack, the lymph tissue of a Peyer's patch resembles a diplomatic center where incoming microbes get the benefit of the doubt— they are presumed "friendly" or at least "noncombatant" until proven otherwise.[32]

To say that the immune system learns to ignore the intestinal bacteria, however, would be wrong. Rather than inducing a microbe-killing inflammation, the interaction on the Peyer's patch triggers the production of an abundance of the antibody known as immunoglobulin A, or IgA. Like all antibodies, each IgA attaches to a specific target, in this case a particular kind of intestinal bacteria. Instead of marking the microbe for destruction (as do most antibodies), IgA simply clusters across its surface to keep it from attaching to the intestinal wall—a gentle "keep moving," as it were.[33] The meet-and-greet inside the neutral territory of the Peyer's patch also leads to the proliferation of T cells and B cells that will marshal an attack against these same bacteria should they turn up in forbidden territory, such as the blood. And so an infant's budding immune system learns to tolerate swallowed bacteria while warily keeping them at a safe distance. To appreciate the crucial nature of this diplomacy, consider what happens when it breaks down. In Crohn's disease and ulcerative colitis, the immune system responds to the touch of harmless intestinal microbes with a frenzy of tissue-killing inflammation, producing excruciating intestinal ulcers that can worsen into deadly perforations.[34]

Over the course of childhood, the number of Peyer's patches lining the small intestine dwindles from several hundred to around thirty. This remnant group of Peyer's patches clusters along the final segment of the small intestine just before it opens up into the expansive bacterial holding chamber of the colon. Within this remnant, a much-reduced diplomatic corps of immune cells continues to monitor the daily passage of millions of microbes, recognizing the vast majority as normal and worthy of tolerance.

Once they have passed through the forceful contractions and microbe-snagging cells of the small intestine, bacteria enter the settling tank of the large intestine. Though it is sterile at birth, it will become the microbial rain forest of the human body. In 1905 the French microbiologist Henri Tissier became the first to study how this jungle of an ecosystem takes shape, as reflected by what comes out the other end.

At the end of a vaginally delivered baby's first day of life, the scattering of bacteria in his poop reflects that of his mother's vaginal and intestinal tract. By contrast, that of the cesarean-born baby contains a more random assortment of microbes from the hands of the birth attendants and the general hospital environment.[35] Whatever the method of delivery, by the third day a breast-fed infant is excreting a near-monoculture of bifidobacteria, which continues to predominate until the introduction of solid food. The intestinal flora of the formula-fed infant also contains bifidobacteria (source unknown) but in far smaller numbers and as a small part of an unstable mixture of other microbes.

In the intestine, as on the baby's skin and in its mouth, the bifidobacteria discourage the growth of potential troublemakers such as staph and help select the first permanent residents. Studies also show that an abundance of intestinal bifidobacteria boosts the level of protective antibodies in a baby's blood—antibodies that target not only problematic bacteria but many kinds of diarrhea-inducing gastrointestinal viruses.[36] This phenomenon may help explain why, in Third World countries with poor water sanitation, the mortality rates of breast-fed infants are as much as six times lower than that of formula-fed babies.[37] Even in the United States, where infant deaths from diarrheal disease remain rare, epidemiologists find as much as a 20 percent higher survival rate in breast-fed versus formula-fed babies during the first six months of life, regardless of family income or education level.[38]

The first wave of intestinal microbes also triggers the maturation of the colon's lining. Underlying blood vessels extend to the lining's surface and there form the dense network of tiny capillaries needed both to keep it healthy and to carry away the nutrients liberated by resident bacteria.[39] At the same time, the first touch of bacteria awakens millions of intestinal stem cells. Once activated, these cells begin endlessly dividing, and their proliferation continually refreshes the intestinal lining's delicate layer of surface cells. The surface cells, in turn, begin shedding at a rate of several billion cells per day. This ceaseless replacement renders the intestinal tract resilient to the kind of injuries that inevitably occur when a child starts eating solid food, with its abundance of natural toxins as well as the occasional sharp object or disease-causing microbe.

The introduction of solid food brings the breast-fed infant's intestinal community more or less in line with that of her formula-fed peers, though no two people end up harboring exactly the same species and strains. On average, some thirty species tend to predominate, with a hundred or so present in smaller numbers. The most abundant and productive of these include fiber-digesting anaerobes such as the rod-shaped *Bacteroides* and *Eubacteria*. Of these, *Bacteroides* such as *B. theta*, *B. vulgatus*, and *B. fragilis* make up 20 to 30 percent of a person's intestinal bacteria and produce around a quart of odorless carbon dioxide and hydrogen gas each day. The *Eubacteria* make themselves known by their production of the more odiferous hydrogen sulfide—familiar the world over for its rotten-egg smell.

In addition, we end up hosting an assortment of "cocci," or orb-shaped anaerobes. They include *Enterococcus*, *Peptococcus*, *Streptococcus*, and *Peptostreptococcus* species, which ferment an assortment of complex proteins and fats (glycoproteins and glycolipids) into simpler sugars and fatty acids that the body can absorb. In the process, they produce another signature odor—the rotten-butter scent of butyric acid.

Together the *Bacteroides*, *Eubacteria*, and various cocci liberate as much as 30 percent of the calories a person absorbs from food, especially from high-carbohydrate meals like bowls of cereal or pasta.[40]

Other major members of the microbiota include the clostridia, some of which produce toxins and all of which can retreat into resistant spores. The most infamous, *Clostridium difficile*, has a nasty tendency to cause diarrhea and colon inflammation after a course of antibiotics has razed its competition. Oddly, most infants harbor *C. difficile* without suffering any ill effects. By adulthood, when this bug can become problematic, the normal microflora keep it in tight check.

Minority members of the intestinal community include a half dozen kinds of lactobacilli and a scattering of facultative anaerobes such as *E. coli*, whose ability to survive in open air allows them to stray and occasionally cause problems in places like a woman's urinary tract. In addition, around one in five of us hosts a detectable amount of methane-producing intestinal bugs such as *Methanobrevibacter smithii* and *Methanosphaera stadtmanae*. These methanogens feed on the hydrogen and carbon dioxide produced by their fiber-digesting neighbors. Like hydrogen, methane proves odorless but flammable, the latter quality one that continues to delight adolescent boys the world over.

The skin. The mouth. The nose. The digestive tract. By the 1980s, microbiologists could boast a basic understanding of what lived where on the human body, as well as a budding appreciation for both their benefits and their dangers. What science could not yet explain was how the immune system tolerated their presence—especially that of the teeming nation within the bowels. Equally mysterious: exactly how our resident bacteria trigger the profound changes that their presence clearly produces in our cells and tissues. Of more than academic interest, the answers promised to reveal how a breakdown in the intricate negotiations between microbe and man can result in disease.

WHO'S THE BOSS?

Gastroenterologist turned gut microbiologist Jeffrey Gordon heads Washington University's gleaming new Center for Genome Sciences, in St. Louis. The expansive, sun-streaked laboratory sits above the school's renowned gene-sequencing center, a major player in the Human Genome Project, which in 2003 completed the sequencing of the 20,000 to 25,000 genes that spell *Homo sapiens*.

"Now it's time to take a broader view of the human genome," says Gordon, "one that recognizes that the human body probably contains a hundred times more microbial genes than human ones." In 2005 Gordon and his collaborators at California's Stanford University and Maryland's Institute for Genome Research pulled together a set of multimillion-dollar grants from private foundations and government agencies to fund their Human Gut Microbiome Initiative, a plan to isolate, sequence, and analyze the sum total of microbial genes that contribute to the health and maintenance of the human body—and occasionally to its malfunction. One aspect of this gargantuan project is to produce a community profile of the intestinal microflora's genetic abilities. Another is to sequence the complete genomes for a hundred of the most representative of the human colon's bacterial residents.

Gordon's "second human genome project" is but one of a dozen-odd research projects unfolding in his lab at any one time—all directed at understanding the influence of intestinal microbes on human health and disease, not only in the bowels but also elsewhere in the body. His laboratory staff consists of a constantly evolving stable of twenty-plus graduate students and postdoctoral fellows with expertise in disciplines ranging from bacterial ecology to X-ray crystallography.

Gordon's interest in colonic bacteria traces back to his years as a gastroenterologist in the 1970s and 1980s, when he studied the human genes that control the cell division that continually refreshes the intestinal lining. This constant replacement (the cells shed when they're just three days old) not only renders the intestinal lining resistant to injury, it also ensures that the colon's resident bacteria neither settle in too deeply nor grow too abundant; the vast majority get swept away aboard discarded epithelial cells with each day's bowel movement.

Gordon appreciated that all this cell division comes at a high cost: the birth of every new cell presents the risk that a random mutation will remove the brakes on cell division and produce cancer. To a gastroenterologist, it comes as no surprise that colon cancer ranks as the second leading cause of cancer death in the industrialized world (behind smoking-related lung cancer).[41]

Gordon's early studies teased out how different cell genes turn on and off at particular times in an intestinal cell's development, at the same time as the cell moves up from crevice to peak along the intestinal villi. He concluded that the genes were receiving exacting instructions. But from where? Conventional thinking would have directed Gordon to look for these biochemical signals in the tissues and organs underlying the intestinal lining. Instead, Gordon became intrigued with the possibility that the cells were receiving their marching orders from the bacteria clinging to their outer surface.

With hundreds of different kinds of bacteria and other microbes living inside the colon at any one time, Gordon knew he needed a simplified model to test his theory. From University of Illinois microbiologist Abigail Salyers, Gordon's staff learned the art of raising germ-free mice. Using them, he could then follow what happened when he added back one member of the microbiota at a time. As the lab's informal mentor, Salyers also gave Gordon a starter set of B. *theta* isolated from the stool of a healthy human volunteer. B. *theta*, she had found, grew as well in the intestines of mice as in those of man. Gordon went on to show it to be a particularly bossy tenant.

For example, his team caught B. *theta* begging for handouts when a mouse missed its usual chow. Salyers had already shown that B. *theta* survived such lean times by feeding on a sugary substance (fucose) secreted by intestinal cells.[42] Gordon's lab then showed that the intestinal cells produced this treat only at B. *theta*'s insistence.[43] First they demonstrated that the intestinal cells of germ-free mice stop making fucose within a few weeks of birth. "It was as if they were preparing for guests who never arrived," says Gordon. But all it takes is a squirt of *Bacteroides* down the gullet of an adult germ-free mouse, and the sugar production immediately resumes. Next, his staff used three types of B. *theta* mutants (again supplied by Salyers) to decipher what exactly was going on. In one set of germ-free mice, they introduced a strain of B. *theta* that couldn't attach directly to the intestinal cells. Regardless, the cells began producing fucose. They colonized another set of germ-free mice with *B. theta* mutants that couldn't consume the fucose. Still, the intestinal cells produced fucose. The food train failed to start only in mice that had been monocolonized with a *B. theta* strain unable to secrete the particular protein that the researchers suspected was a biochemical begging signal. In other words, the cells' sugar production arose not merely in response to the touch of bacteria or the depletion of fucose. The key was a "feed me" message from *B. theta* that turned on a mouse gene that otherwise became dormant after the first few days of life. The discovery was Gordon's first clear confirmation of his once-outlandish idea that intestinal bacteria can directly control the activities of intestinal cells.

In the 1990s, the advent of DNA microarrays, or gene chips, gave Gordon a powerful new tool for his research. Gene chips allow scientists to scan for the activity of thousands of genes at once. They use thousands of fluorescently labeled snippets of DNA that have been assembled at precise locations across a grid the size of a microscope slide.[44] In 2002 Gordon's lab used a "Mouse Chip" containing some twenty thousand known mouse genes to document that hundreds of these genes get switched on when a previously germ-free mouse gets its first infusion of *B. theta*.[45] As Gordon expected, these genes included many of those involved with the normal maturation of the intestinal lining. Introduction of *B. theta* also switched on mouse genes involved in the production of the specific transport molecules needed for intestinal cells to absorb and use the many nutrients supplied to them by *B. theta* and related bacteria.[46] All this reinforced Gordon's growing impression that *B. theta* plays a particularly important role in promoting intestinal health.

Gordon's crew completed its sequencing of *B. theta*'s 4,779 protein-making genes in 2003, the same year as the Human Genome Project completed its sequencing of *Homo sapiens*.[47] They found that *B. theta* dedicated more than one hundred of these genes to retrieving undigested plant sugars, and another 170 to breaking them down into components that a mouse (or human) could absorb. *B. theta* also possessed an elaborate apparatus for sensing what nutrients were available to it at any given time, so that it could assemble the correct combination of biochemical tools to deal with them.

The lab's sequencing of *B. theta*'s genome also gave Gordon's researchers a "*B. theta* on a chip" to complement their mouse micro-array. Now they could follow both sides of the biochemical conversation taking place between host and microbe. The following year, 2004, the lab discovered that *B. theta*'s bossiness extended beyond the intestines. They caught it issuing marching orders to fat cells in the mouse's abdomen.[48] Specifically, they found that *B. theta* stopped the production of a fat-suppressing hormone known as fasting-induced adipocyte factor, or Fiaf. This discovery largely explained an earlier observation. When the lab colonized germ-free mice with squirts of *B. theta* down their throats, the animals immediately began laying down abdominal fat, even while eating 30 percent less chow and undergoing a spike in metabolic rate that burned almost 30 percent more calories. After fourteen days with *B. theta*, the mice had increased their fat stores by an average of 60 percent.

"Here we are watching *B. theta* exert a hormone-like effect on its host," marvels Gordon. "It's as if *B. theta* is saying, 'Save this—we may need it later.'"

Further exploring the complexity of such symbiotic interactions, Gordon's team has shown that early in life, an animal's intestinal cells and immune system begin producing substances that help beneficial bacteria such as *B. theta* anchor themselves in place while other, potentially dangerous microbes get flushed out of the colon.[49] *B. theta* appears to return the favor by not taking undue advantage of its host. For example, these bacteria wait until the intestinal lining sheds its sugar-coated epithelial cells before they begin grazing on them. And they don't beg for sugar unless their normal supply of undigested plant matter disappears.

Through it all, *B. theta* imparts a kind of stability to its ecosystem, Gordon concludes. The bacteria turn to their host when outside food is scarce, and in good times they provide their host with additional calories and with instructions to squirrel at least some of this extra bounty away for hard times ahead. Most recently Gordon and his crew have turned up evidence that a second large group of intestinal bacteria—the Firmicutes—may be even more efficient at extracting and sharing calories than are the Bacteroidetes (of which *B. theta* is a chief member).[50] "While an overfed nation may not appreciate the extra calories and fat," says Gordon, "I imagine that there were great periods in our human experience when this kind of dynamic spelled the difference between starvation and survival." Not surprisingly, Gordon's discoveries have sparked interest in the possibility that tinkering with our intestinal microflora might help the obese lose weight.

As further evidence, still other lab members have described what happens when a mouse monocolonized with *B. theta* gets an infusion of the methane-producing microbe *Methanobrevibacter smithii*: the double-colonized mice end up hosting one hundred times more *B. theta* than they otherwise would. As it turns out, methanogens such as *M. smithii* greatly increase *B. theta*'s efficiency by feeding on its waste products of hydrogen and carbon dioxide and converting them into methane and water. Without the methanogens, the accumulating wastes would slow down *B. theta*'s simple metabolism and limit its ability to multiply.[51] On a practical level, the increased efficiency produces an additional 15 percent fat gain in the co-colonized mice.

Lab member Ruth Ley has also launched a major exploration of the similarities and differences in the microflora of man and beast—this to better understand the evolutionary roots of our nation within. Of fifty-five known divisions of bacteria on planet Earth, only eight make their home inside the digestive tract of animals, she points out, which suggests this is a highly selective relationship. "Our hypothesis," she says, "is that they have coevolved with us over millions of years."

On a recent afternoon, Ley stands over a bucket of ice packed with vials of St. Louis Zoo doo—cheetah, lion, elephant, kangaroo, hyena—as well as a sampling of dung collected by a colleague doing research near an African waterhole. "If it's true that we, as mammals, coevolved with our intestinal microflora, then we should see similarities suggesting that some ancient bacterium got inside some ancient ancestor of all these species and set up shop." So far Ley has found many broad similarities. While eight divisions of bacteria appear in the intestinal tracts of mammals (including humans), just three of those divisions— Bacteroidetes, Firmicutes, and Proteobacteria—predominate.

By contrast, an abundance of different groups crop up from mammal to mammal when Ley parses bacteria on the level of genus (the grouping just above species). *Bacteroides* such as *B. theta*, *B. vulgatus*, and *B. distasonis* predominate in omnivores—creatures such as ourselves, mice, and pigs that eat both plants and meat. In herbivores such as cattle, sheep, and rabbits, the number-one spot goes to members of the closely related *Prevotella* genus (*P. ruminicola*, *P. brevis*, *P. albensis*, et al.). They are the kinds of differences an evolutionary biologist would expect to appear as new species of mammals and their resident microbes branched off in ancient times to pursue their different lifestyles.

A NEW WINDOW OPENS

The anaerobic culture techniques developed by Holdeman and Moore, together with Gordon's genetic eavesdropping, have made the microflora of the large intestine the best understood of the body's many microbial ecosystems. But at least 10 percent of the intestinal tract's resident species remain uncultured and undescribed. In the dawning years of the twenty-first century another revolutionary technology has gone far in revealing the last of these, our most mysterious microbial residents. In doing so it has also confounded the world of medicine by detecting bacteria in parts of the body that were thought microbe-free except when seriously diseased.

This technology—the bacterial gene probe—emerged from the work of University of Illinois microbiologist Carl Woese. In the 1970s and 1980s, Woese was searching for a genetic yardstick to indicate relatedness among the Earth's bacteria. Scientists had long been searching for a better way to arrange organisms into family groups than by clustering them according to superficial traits such as appearance and function—a tactic that risked grouping the microbial equivalent of butterflies with bats. Because all genes accumulate tiny inconsequential changes over time, Woese knew he needed a gene that was both crucial to all living cells and complex enough that subtle variations within its DNA letters could be used to measure evolutionary distance. For his yardstick Woese chose a gene coding for a crucial segment of a bacterial ribosome, or protein factory. He thereby discovered a wholly unexpected split in the tree of life—an early separation that produced a distinct lineage of ancient bacteria-like microbes that he dubbed archaea. The genetically distinct archaea turned up primarily in extreme environments such as deep-sea vents and sulfur hot springs, but they included a few human residents such as the methane-producing *Methanobrevibacters*.[52]

While Woese was redrawing the tree of life, one of his postdoctoral fellows, Norman Pace, realized that the same sort of signature gene could be used as a kind of DNA fingerprint to identify the multitude of bacteria in an environmental sample such as a scoop of soil or water. That is, he could design a DNA probe to search for bacterial ribosomal genes in soil or water, using as his target the portion of the gene that is the same in all bacteria. Pulling these genes out of his sample, he could then copy the segments thousands of times over with PCR, or polymerase chain reaction—the same gene-amplifying technique used in forensic labs to amplify the genetic "fingerprints" left at crime scenes. Pace could then sort his gene segments according to their subtle differences. The beauty of the method was that it enabled Pace to identify bacteria in a mixed sample by using the DNA letters of a single gene as a "bar code" — this being vastly simpler than isolating and growing out the individual species in the lab to identify them by differences in chemistry and appearance.[53]

By the end of the 1980s, microbiologists the world over had enthusiastically embraced Pace's new tool. In particular, they had settled on one ribosomal RNA gene—that for a segment of the ribosome dubbed 16S rRNA—as their DNA fingerprint of choice. It rapidly became routine to sequence this gene for any bacterium under study.[54] The result was an ever-growing library of 16S genes that could be used to identify bacteria in the same way that forensic investigators use genetic databases to match crime-scene DNA to that of known criminals.

Most profoundly, perhaps, 16S gene probes provided the first direct method for identifying bacteria that were impossible to grow in pure culture—that is, separate from the confusing jumble of other microbes that help knit together their natural communities.[55] And if a particular rRNA signature does *not* show up in any library of known microbes, eureka! You've discovered a new species. Even better, you can place this new species in a general family group, perhaps even a genus, by searching for its closest match among well-described species.

Pace's bacterial gene probes opened up a new world of discovery for microbiologists. In 1986, for example, he reported that previously unknown and uncultivated bacteria made up an unbelievable 99 percent of the bacteria in many of his soil, mud, and water samples.[56] Might the same prove true of the complex microbial communities inhabiting the human body?

Among the first to begin fishing 16S rRNA genes from human tissues were the microbiologists David Relman of Stanford University and Ken Wilson of Duke University. In 1991, working independently of each other, they both turned their probes on the tissues of AIDS patients suffering from Whipple's disease. Rare outside of the immune compromised, the disorder is marked by severe weight loss, arthritis, and organ damage.[57] For eighty-five years medical researchers had failed to culture the small rod-shaped organism they glimpsed in the tissues of their patients, and so they had no way to identify it or compare it to other, known kinds of bacteria. Wilson and Relman's 16S gene sequences of the mystery bug placed it firmly in the order Actinomycetales, bacteria that form branching fungus-like filaments; this group includes many members of the mouth and intestinal microflora. More important, they gave the medical community a novel genetic test that enabled quick diagnosis—vital because prompt treatment of Whipple's disease, with an appropriate antibiotic, stops the infection before it can permanently damage the heart and brain.

A couple of years later Relman went 16S fishing again, this time in a healthy human mouth (his own), for the sole purpose of plumbing its diversity. He came back to the lab from a dental appointment with several presterilized tubes containing tooth scrapings taken just below the gum line. He and several of his lab members then amplified the bacterial DNA in the samples and sequenced 264 distinct 16S rRNA genes. Just over half matched the 16S genes of known bacteria, and thirty-five proved different enough to indicate entirely new species (rather than strains, or subtypes, of known ones).[58]

Since 1994, Bruce Paster and Floyd Dewhirst, of the Harvard-affiliated Forsyth Institute, have been using 16S gene probes to take far more comprehensive surveys of mouth microbiota, plumbing the diversity of organisms in dozens of volunteers, some healthy and others suffering from various oral ills. So far they've discovered the genetic signatures of more than seven hundred kinds of oral bacteria, the majority of them previously unknown. A typical mouth harbors one to two hundred of these organisms, Paster and Dewhirst found; some of them are consistently associated with problems, while others promote sweet breath and good oral health.[59]

The Forsyth scientists have used their extensive database of 16S genes to build a DNA microarray that can be used as a kind of bar code reader to quickly identify which of more than two hundred kinds of predominant oral bacteria a person has in his or her mouth and whether they are present in relatively low or high amounts. The microarray, though not yet ready for use in the dental office, allows researchers to assess a person's risk of oral disease by seeing what "good guys" are missing as well as what troublemakers have crept in. It also enables them to monitor what happens to the complex consortia of bacteria in a person's mouth with treatment—be it a root canal or antibiotics, both of which sometimes create new problems.

DNA microarrays have allowed microbiologists to make similar surveys of other body niches. In 2006, for example, Relman and Gordon joined scientists at the Institute for Genome Research, in Rockville, Maryland, to complete the ambitious 16S sequencing of the microbial community inside the intestinal tract. They turned up more than two thousand different 16S genes in the stool samples of two healthy adults, around 150 of these different enough to be distinct kinds of bacteria and thirty-five of them new to science.[60]

Meanwhile, Relman and others have been turning up bacteria in tissues that were long thought microbe-free. "It may turn out to be rare to find a human tissue that *doesn't* show the presence of bacterial DNA," says Relman.[61] Whether those microbes represent hidden infections or normal microflora remains to be seen. "What I hope," he says, "is that by starting with specimens from healthy people, the assumption would be that these microbes have probably been with us for some time relative to our stay on this planet and may, in fact, be important to our health."

STEALTH INFECTIONS OR INNOCENT BYSTANDERS?

Admittedly, the behavior of even well-known bacterial inhabitants is challenging the old, straightforward view of infectious disease as embodied in Koch's sacrosanct postulates—namely, that any microbe that causes a disease should turn up in every case of the disease and should always cause the disease when introduced into a new host. *Helicobacter pylori* may be the most familiar germ to defy this rubric. Once a near-universal resident of the human stomach, only in modern times did it begin causing the gastric ulcers. Even today *H. pylori* proves troublesome in only a small minority of those who carry it.

"This stuff drives the old-time microbiologists mad," says molecular biologist Alan Hudson, "because Koch's postulates simply don't apply." Instead, the kinds of hidden infections typified by *H. pylori* and other newly discovered microbes appear to cause problems only in some people and only sometimes, typically after years if not decades of uneasy peace between host and bacterium.

Hudson, a microbiologist who takes pride in never having owned a microscope, enjoys telling a personal anecdote that well illustrates how gene probes have not only revolutionized but confounded the science of medical microbiology.[62] "A young soldier gets chlamydia, a genital infection, overseas, clears it up with antibiotics, and comes home and marries his high school girlfriend," Hudson begins, stretching back from his desk in a crowded corner of his laboratory at Wayne State University in Detroit. "But three weeks later they both have screaming chlamydia infections."

Acting out the double standard, the husband is outraged that his girl has been screwing around while he was away—a charge she flatly denies. "So I get a call from the small-town doctor who's trying to save their marriage," says Hudson. "He's known her since childhood and believes she's telling the truth." The doc wants to know whether there's any way the husband could have infected her, though he appeared to have been properly treated and cured. "So I tell him, 'Send me a urine specimen from him and a cervical swab from her.'" This after both had completed a full course of antibiotics under the doctor's supervision. "I PCR'd them both," Hudson says, "and he was still red-hot." Instead of clearing up the chlamydia, Hudson theorizes, the antibiotics simply drove it into a dormant state. If one of those quiescent cells then passed into a new host—the bride—it may have begun dividing to create an active infection that she then passed back to him.

Though dormant chlamydia rarely reactivates in a way that produces a new genital infection, Hudson has long suspected it of other mischief. In the early 1990s, while analyzing the joint tissue of arthritis sufferers, Hudson turned up two types of chlamydia—*Chlamydia trachomatis*, normally associated with genital and eye infections, and *Chlamydia pneumoniae*, a common cause of respiratory infections.[63] More famously, in 1996 he began fishing *C. pneumoniae* out of the brain cells of Alzheimer's victims.[64]

Around the same time, medical researchers began finding the genetic fingerprints of *C. pneumoniae* and various kinds of mouth bacteria in the arterial plaque of heart attack patients.[65] This finding spurred many cardiologists to begin putting their patients on antibiotics, a practice largely stopped in 2005. That year brought the much-anticipated results of a study involving more than four thousand heart disease patients who had been taking the powerful antibiotic gatifloxacin for two years. The results of the study confirmed that the treatment failed to reduce the risk of heart attack or the degree of atherosclerosis, or artery blockage.[66] However, the trial did not so much exculpate *C. pneumoniae* as demonstrate that even long-term courses of strong antibiotics fail to fully eradicate it. So the study, while it squelched the growing use of gatifloxacin among cardiologists, opened up tremendous interest in developing stronger, more effective drug regimens to eradicate not only *C. pneumoniae* but also a growing number of other so-called stealth infections.

Clearly, the financial incentives are huge. If, for example, a powerful new antibiotic proved even partially effective in reducing the risk of heart attack, the resulting prescriptions would number in the millions, possibly tens of millions. "We're talking about the majority of the population being on long-term antibiotics, possibly multiple antibiotics," says Vanderbilt University chlamydia specialist William Mitchell, cofounder of a company pursuing just such a cure.[67] Moreover, if preliminary results are any indication, these would be prescriptions not for a several-day course of antibiotic, but for months, possibly years, of daily use.

Already, many rheumatologists prescribe long-term—even lifelong—courses of antibiotics for inflammatory arthritis because the drugs reduce painful inflammation.[68] (It's not clear whether they do so by clearing away stealthy bacteria or by some other unknown process.) More recently, psychiatrists have begun putting young patients with obsessive-compulsive disorder on long-term antibiotics, a trend that began with the idea that these children may be suffering from a kind of neurological autoimmune reaction triggered by stray *Streptococcus pyogenes* bacteria left behind from an active infection.[69]

The cataloging of our body's microflora, the ability to raise germ-free animal hosts, and the development of bacterial gene probes have all led to incremental advances in our understanding of the interactions—for better and for worse—between human beings and their microbial hitchhikers. For all anyone knows, it's perfectly normal for these microbes to stray throughout our bodies on occasion—with genetic technology only now revealing their presence.

Hudson, for one, cautions that before we set out to eradicate our bacterial fellow travelers, "we'd damn well better understand what they're doing in there." To that end he has begun working with his own set of DNA microarrays—specifically one gene chip for chlamydia and a second for its human host. Together they are enabling him to eavesdrop on the biochemical conversations taking place between the body and the semidormant chlamydia he's found lingering in joint tissues. He has even resorted to the device he long shunned in favor of DNA probes: a microscope, albeit a $250,000 digitized light microscope that can magnify live organisms an unprecedented fifteen thousand times. The scope sits in the laboratory of Hudson's spouse, Judith Whittum-Hudson, a Wayne State immunologist who is working on a chlamydia vaccine.

Through the scope's video screen he has watched chlamydia cells morph from their infectious, active stage into their little-understood persistent form. "First you have this perfectly normal, spherical bacterium, and then you end up with this big, goofy-looking doofus of a microbe," he says. On a recent spring afternoon he leans closer to the scope's video screen as he focuses its lens on a roiling spot of activity inside a chlamydia cell. "It's doing something," he says. "It's making something. It's saying something to its host."

PART III

TOO CLEAN?

The love of dirt is among the earliest of passions, as it is the latest. Mud-pies gratify one of our first and best instincts. So long as we are dirty, we are pure.
                                        —CHARLES DUDLEY WARNER, 1870

Rohan Kremer Guha—a doe-eyed New Jersey boy with a shy smile and soft black hair— knows not to touch the crumbs and dribbles that other kids leave behind at birthday parties. "I worry," says his mother, Devyani Guha. "What if he were to touch an ice cream stain? But I don't want him to stand out so much. So I let him go and try not to hover." Admittedly, Rohan knows of plenty of other kids with allergies, though none are quite as wide-ranging as his.

Rohan's allergic woes began early. By six months his once-beautiful baby skin had become red and scaly with eczema. And whenever the family's cleaning ladies came to the house, Devyani or the babysitter had to take Rohan outside for the day. "Otherwise, every time the carpet got vacuumed, he'd get a terrible rash and start wheezing." As a toddler, Rohan tested allergic to virtually every kind of food on the roster. The worst offenders—milk, eggs, and wheat—provoked immediate nausea and vomiting. Then, when Rohan was two, a kiss on the cheek from his grandfather produced quarter-sized welts over half the boy's face. A half hour earlier, Grandpa had eaten a mouthful of kaju-katli, an East Indian fudge containing butter and cashews.

By this time Rohan's mother was feeding him little more than chicken and mashed potatoes and had put herself on the same restrictive diet so she could continue to breast-feed. Even the most hypoallergenic formula had caused problems, and of course, cow's milk was out of the question. "I was starving, and we were worried about his nutrition," Devyani recalls. So she was hopeful when, in January 2001, Rohan's allergist recommended she bring him into the hospital for a battery of food challenges. "He told us that a lot of children who test mildly positive to everything in reality are not that allergic to many foods." Just to be safe, the hospital would have a pediatric crash cart on hand. The wheeled cabinet contained everything needed to pull a child back from respiratory failure and cardiac arrest.

At ten a.m. on a weekday morning a hungry two-and-a-half-year-old Rohan sat on a hospital bed, his thin almond-brown limbs lost in an oversize cotton gown. A nurse had already started a drip of ordinary saline solution into his right arm—another "just in case" measure that would allow the medical team to quickly infuse him with lifesaving drugs. A nurse practitioner brought a tray holding a dozen number-coded but otherwise unmarked containers of rice pudding. Into each one a dietitian had hidden an eighth-teaspoon of a food to which Rohan might be sensitive.

Rohan downed the first three samples without complaint but balked after the tiniest taste of the fourth. He began fussing and complaining that his tummy hurt. His mother was trying to coax his attention back to the spoonful of rice pudding when he threw up, then passed out. The nurse practitioner immediately injected him with epinephrine and, within seconds, started an infusion of antihistamine through his intravenous line. When Rohan came back to consciousness, the nurse sat him up in bed and strapped a mask to his face through which he could inhale immune-calming steroids. "Scary," says Devyani. "That's when it really hit that we were dealing with something that could kill our son." Hidden in the fourth sample had been a minute amount of barley.

Six months later, Rohan's new baby brother, Zubin, likewise began developing severe allergies. Devyani quit her job as a successful urban planner specializing in bringing health care clinics to inner-city neighborhoods. "I didn't feel it was safe for me to go out, both in terms of leaving the children and in terms of how careful I have to be about what I eat and pass to Zubin through my breast milk." At ages eight and four, both boys have outgrown some of their food allergies. Soy and chickpeas are back on the family menu. But the boys remain prone to severe asthma and eczema. As is often the case, their asthma attacks tend to be particularly severe and frightening after respiratory infections such as the flu. But Devyani can't get the boys immunized because the flu vaccine contains eggs, one of the foods to which they remain severely allergic.

Throughout the developed world, allergies, asthma, and other types of inflammatory disorders have gone from virtually unknown to commonplace in modern times.[1] All involve a destructive immune system response to a harmless substance such as a food, pollen, the normal bacteria of the colon, or even the body's own healthy cells. An immune attack on the latter can result in autoimmune disorders such as type 1 diabetes, multiple sclerosis, lupus, and many others. It's as if modern man's immune system has lost some sort of safety stop, leaving it on hair trigger, perhaps unable to tell friend from foe. How this over-zealousness manifests itself seems to depend on a person's genetic pre-disposition. Allergies and asthma clearly run in families, as may autoimmunity. But the dramatic increases in the prevalence of these conditions over the last 150 years have been far too rapid to be explained by family genetics. They remain rare in Third World countries, and the children of immigrants tend to share the higher disease rates of their Western peers, as do their parents to a lesser degree. In fact, many African and eastern European newcomers to this country half-jokingly refer to allergies as "citizenship disease," as the signature sniffling and sneezing often shows up right around the time they reach the five-year mark that qualfies them for citizenship.[2]

Is it caused by air pollution? The stresses of modern living? An over-abundance of food? An underabundance of childhood disease? All these things have been proposed as the culprits in the historic rise in immune disorders in the industrialized world. But studies have largely failed to turn up supporting evidence.

Perhaps the strangest suggestion of all has been that this epidemic is the result of our relatively sudden separation from the sea of microbes, largely harmless, that once imbued our lives through the untreated water we drank, the food we pulled from the soil and crudely stored, the animals we raised and hunted, and the dirt on which we walked, worked, and often slept. A truism of biology is that evolution turns the unavoidable into the essential. Could it be that the human immune system evolved in such a way that it came to rely on constant exposure to bacteria and other "germs" to function properly? If so, what happens when over one or two centuries, an evolutionary blink of an eye, we separate ourselves from this continual exposure by sanitizing our water, processing our food, dousing our bodies with germ-killing drugs and soaps, and distancing ourselves from the natural landscape?

FROM HIPPOCRATES TO THE HYGIENE HYPOTHESIS

A scan of medical history reveals a similar pattern of increase for the three categories of inflammatory disorders that have become so prominent in the twenty-first century. They include allergies and asthma; autoimmune disorders such as type 1 diabetes, multiple sclerosis, and lupus; and inflammatory bowel diseases such as Crohn's and ulcerative colitis. All were rare to nonexistent in the ancient medical literature. Hippocrates knew asthma, for example, but it was a disease triggered by exercise, not by allergies. He also describes rare individuals who have bad reactions to certain foods such as milk, but the symptoms—upset stomach and gas—sound more like digestive intolerance than allergy.[3]

The first mention of respiratory allergies dates to the early tenth century and Persian physician Al-Razi's treatise "On the Reason Why the Heads of People Swell at the Times of Roses and Produce Catarrh."[4] The rose cold, or seasonal allergy, next appeared in European medical literature in the seventeenth century and remained a medical curiosity until the nineteenth century, when it became a fashionable malady among the aristocracy, many of whom, it was written, "cannot bear the country air."[5] In 1819 the London physician John Bostock began a decade-long investigation into what he renamed *catarrhus aestivus*, or the summer catarrh (stuffy nose). "One of the most remarkable circumstances respecting this complaint is its not having been noticed as a specific affection until within the last ten to twelve years," he wrote. In 1828 he presented his account of twenty-eight cases of summer catarrh to London's Medical and Chirurgical Society, reporting that the disorder occurred only "in the middle or upper classes of society, some indeed of high rank." Having made inquiries of pharmacies in the poorer sections of London and elsewhere, Bostock likewise reported that he found "not a single unequivocal case occurring among the poor."[6]

In 1873 the Manchester physician Charles Blackley, himself a hay fever sufferer, showed that the condition was a respiratory allergy by storing summer grass pollen in a bottle, opening the stopper in winter, and taking a whiff. It instantly triggered his usual symptoms of streaming eyes, runny nose, and sneezing. In announcing his results, Blackley, like Bostock before him, remarked on the dramatic increase in the condition over the previous decades and expressed puzzlement over the fact that it didn't exist among farm families, who had the greatest exposure to pollen. Blackley discounted the popular idea that respiratory allergies were some type of inbred affliction associated with royal blood, for they were common among the nouveaux riches of European industry and trade. Rather, Blackley described it as a disease of "the educated classes" and suggested the presence of some kind of "predisposition which mental culture generates."[7]

By the beginning of the twentieth century, respiratory allergies had become so commonplace across western Europe and North America that many cities had "hay fever societies" to provide support, or at least commiseration, for the suffering. In 1988 medical historian Martin Emanuel dubbed respiratory allergies a "post-industrial revolution epidemic," affecting around 10 percent of the U.S. population. "Nevertheless," he wrote, "the reason for this increase is understood no better today than when Blackley noted it in 1973."[8] Add in food allergies, also rare until the twentieth century, and allergic disorders today plague nearly 60 million Americans, or 20 percent of the population. As many as 15 million of these people suffer allergies severe enough to trigger life-threatening anaphylaxis.[9]

The prevalence of asthma followed in lockstep with that of respiratory allergies, even as it changed from the exercise-induced disease familiar to the ancient Greeks to one triggered primarily by allergens, especially indoor irritants such as mold and the microscopic feces of dust mites and cockroaches.[10] The first clear reference to this type of allergy-induced asthma dates to 1552, when the celebrated Renaissance physician Girolamo Cardano traveled to Scotland and cured the Archbishop of St. Andrews of his chronic breathing problems by getting rid of his presumably mite-infested feather pillows and quilt.[11] Over the next four centuries, asthma shared hay fever's association with the upper classes, the classic image of the asthmatic being the pale and pampered young child who eschewed the outdoors. By the 1980s asthma had become the most common chronic disease of childhood and the leading cause of school absences and childhood hospitalizations in North America and Europe, especially in cities.[12]

By the mid-twentieth century it had become clear that a tendency toward allergic disorders ran in families. However, the same tendency could manifest as a food allergy in one family member, a respiratory allergy or asthma in another, and an allergic skin condition such as eczema in a third. Having any one of these disorders greatly increased the chances that a person would develop others.[13] This discovery led researchers to speculate on what elements in a modern lifestyle—especially an upper-crust or urban lifestyle—pushed those with a genetic predisposition over the line into disease. The hunt for those factors began in earnest in the 1980s.

The epidemiologist David Strachan, a skinny young Scotsman with black-framed glasses perched on his jutting ears, arrived at London's School of Hygiene and Tropical Medicine as a lecturer in 1987, having published a tidy string of scientific articles on the prevalence of childhood asthma and its relation to the home environment. Eschewing the kind of slog-through-the-Third-World adventures for which London School epidemiologists were famous, Strachan found his niche in the dog-eared files of family physicians and the near-indecipherable notes they scribbled in patient charts. Open bedroom windows and damp housing ranked among the suspected culprits in Strachan's analyses.[14]

At the London School, Strachan quickly expanded his explorations beyond the notations of family doctors and into the deep databases of the United Kingdom's nationalized health care system. Open windows and mildew alone could not account for the nation's growing epidemic of atopy, or allergic disease. In the 1980s an estimated one in eight British children suffered some variant—be it eczema, food allergy, hay fever, or asthma.

Strachan examined the statistical gold mine of the U.K.'s National Child Development Study, a colossal research effort aimed at tracking the health and welfare of 17,414 Britons—every child born between March 3 and March 9, 1958, near the height of the postwar baby boom. This cohort had by now reached their twenties. So Strachan could cross-tabulate and logistically regress the minutiae of their lives from birth to young adulthood to expose anything that might correlate with two common markers of allergic predisposition: eczema and hay fever.

The one striking association that sifted to the top of Strachan's analyses was family size: the more siblings a child had, the lower was his risk of developing eczema or a respiratory allergy. Strachan confirmed his results with the health statistics coming out of a second birth cohort—a week's worth of British children born in the spring of 1970. At five years of age, they too showed an inverse relationship between family size and the risk of allergy.

Strachan's statistics didn't tell him what it was, exactly, about an abundance of siblings that protected a child from developing allergies. But he felt sure of the answer. In his 1989 report in the *British Medical Journal*, he concluded, "Over the past century, declining family size, improvements in household amenities, and higher standards of personal cleanliness have reduced the opportunity for cross infection in young families. This may have resulted in more widespread clinical expression of atopic disease, emerging earlier in wealthier people."[15] The modern epidemic of allergies and asthma, Strachan concluded, stemmed directly from a decrease in the usual viral infections of childhood, from the common cold to measles, mumps, and rubella.

Strachan chose the alliterative title "Hay Fever, Hygiene, and Household Size" for his succinct report. Journalists did him one better with the "hygiene hypothesis," as they reveled in his counterintuitive indictment of the spotlessly scrubbed home life epitomized in TV fare of the day. His article gave the growing antivaccine movement strong ammunition: its underlying argument is that by preventing childhood infections, modern medicine robs the immune system of the training needed for lasting health.

In 1999 Norwegian researchers added a twist to Strachan's hygiene hypothesis with their finding that older brothers, rather than sisters, conveyed the bulk of the allergy protection enjoyed by those born into large families. The same study showed additional protection coming with family pets, particularly dogs.[16] Strachan seized on the report as confirmation of his ideas, brothers being widely regarded as germier than their presumably prissier sisters.[17] Still, boys clearly didn't get sicker more often than did girls, and dogs could hardly be blamed for spreading colds, flu, and measles.

Meanwhile, immunologists were weighing in with a possible mechanism to explain the apparent protection conveyed by early childhood infections. Their studies of the immune cells circulating in the blood of the allergy-prone showed an imbalance in two newly discovered subsets of "helper T cells." The immune system's major generals, T cells respond to antigens by secreting a complex mélange of signaling molecules called cytokines. By definition, an antigen is any substance that binds to a T cell's surface receptors and triggers a proliferation of T cell clones targeted specifically against it. For disease-fighting purposes, an antigen consists of an identifying piece of a virus, a bacterium, or a diseased or damaged cell that requires elimination. In allergies, the antigen, or "allergen," is part of a substance that the immune system mistakes as dangerous. In the case of autoimmunity, the antigen may sit on the surface of a specific type of cell that the immune system mistakenly marks for destruction.

Immunologists of the late 1980s discovered a dichotomy in the immune system's T cell response. One type of T cell, dubbed type 1 helper cells, or Th1, issued cytokines that direct the soldier cells of the immune system (macrophages, killer T cells, and the like) to devour infected, cancerous, or otherwise diseased body cells. By contrast, Th2 helper cells secrete cytokines that marshal an immune response marked by a rush of mast cells and basophils to mucous membranes, where they release histamine and otherwise produce inflammation and the contraction of underlying muscles. The latter strategy appears to be designed to flush out intestinal parasites such as tapeworms and nematodes, or at least their larvae, so that they cannot attach themselves and grow into adult worms. Curiously, prolonged infection with intestinal parasites actually suppresses the Th2 response, most likely because a prolonged inflammatory response would do more damage than would the worms. In any case, in the absence of such parasites, an overzealous amount of Th2 cell activity appeared to drive the inflammation, fluid release, and muscle spasms of allergies and asthma.

Further study suggested that newborns come into the world with their immune systems skewed toward a Th2 type of immune response, perhaps because the "normal" Th1 cell-killing response could result in a deadly clash with the mother's tissues and immune system. Normally, the type 2 helper response eases back in the first weeks to months of life. By contrast, in the allergic child Th2 cells and their associated cytokines remain at inappropriately high levels into adulthood.[18] These T cell studies fit nicely with Strachan's hygiene hypothesis, in that a lack of early childhood infections might leave a person's immune system inappropriately stuck in early infancy's Th2 mode.

The paradigm of Th1-Th2 imbalance gave the hygiene hypothesis scientific heft to complement its popular appeal. In 1997 even the pundits at *The Economist* seemed convinced: a lengthy editorial, "Plagued by Cures," invoked the hygiene hypothesis as suggesting "that intervening in infections may have undesirable effects on the hosts—that is, on people—as well as the pathogens themselves."[19] Cementing the hygiene hypothesis into medical dogma, *The New England Journal of Medicine* published a University of Arizona study that followed the health records of more than a thousand children, from birth in the early 1980s through age thirteen. The results confirmed a strong protection against asthma associated with early exposure to lots of children—be they older siblings or day care classmates.[20] Like Strachan, the Arizona pediatricians attributed the protective effect to the abundance of colds and other respiratory infections that children spread among classmates and bring home to younger siblings.

Less publicized were the results of several studies that *failed* to confirm the purported link between early respiratory infections and protection from allergies and asthma. In 1996 Strachan himself was unable to find a direct correlation when he ran a statistical analysis between the number of colds and chest infections that babies experienced during the first year of life and the development of hay fever in later childhood.[21] Still other studies confirmed the protective effect of large families but showed that respiratory infections during infancy actually boosted the chances that a child would develop allergic disorders. Closer analysis showed that the increased risk correlated less with infection per se than with the use of antibiotics.[22] (Most respiratory infections are viral and so don't warrant antibiotics.)

Around the same time, the idea that allergies resulted from a teeter-totter imbalance between the two arms of the immune system crumbled in the face of a simple observation. If allergy and asthma resulted from an immune system skewed too far in the direction of a Th2-type immune response, then the Western world should be enjoying a corresponding decrease in the kinds of disorders that result from an overly aggressive Th1 response: that is, the diseases of autoimmunity, in which the cell-killing arm of the immune system mistakenly destroys healthy tissues. But just the opposite was happening.

A HISTORY OF SELF-DESTRUCTION

As with allergies and asthma, many types of autoimmune disorders went from rare or wholly unknown to relatively common during the nineteenth and twentieth centuries. In 1966 Harvard neurologist David

Poskanzer recognized the "twin sanitation gradients" of latitude and wealth that underlay the peculiar epidemiology of multiple sclerosis. The rising prevalence of MS coincided, more than anything else, with the introduction of indoor plumbing across northwestern Europe and North America. The exception that proved the rule, Poskanzer noted, was the low rate of multiple sclerosis in Japan—a country high on the latitude and wealth gradient but notable for the unsanitary practice of using "night soil," or human sewage, as crop fertilizer.[23] In line with Poskanzer's research, it became popular in the 1970s to blame MS on stay-at-home moms obsessed with maintaining a spotless household.

Epidemiologists found similar wealth gradients for other auto-immune disorders. The most striking was type 1, or insulin-dependent, diabetes, which results when the immune system destroys the insulin-secreting cells of the pancreas.[24] (By contrast, type 2 diabetes is caused by insulin resistance, or the body's failure to respond to normal amounts of insulin, often after years of obesity.) In 2000 University of Leeds pediatric epidemiologist Patricia McKinney reported that the risk of developing type 1 diabetes decreased in direct proportion to the amount of time a child spent in day care during infancy and the number of other children in attendance. Placing babies in day care with more than twenty children cut their risk of developing this type of diabetes in half.[25] And in 2004 Lancaster University pathologist Jim Morris and statistician Amanda Chetwynd showed a similarly strong protection against diabetes among babies who either shared a bedroom with other children or had regular and early contact with pets or farm animals.[26]

Of some eighty different kinds of autoimmune disorders tallied at the end of the twentieth century, many had tripled or quadrupled in prevalence over the previous fifty years, and others had appeared out of nowhere.[27] In addition to MS and diabetes, they included lupus and scleroderma, in which the immune system launches a generalized attack on connective tissue; rheumatoid arthritis, which destroys joint tissue; the muscle-wasting disease myasthenia gravis, with its hallmark destruction of motor-receptor cells; Addison's disease, in which the adrenal cells come under fire; and Hashimoto's, marked by thyroid destruction. All together, autoimmune disorders now affect somewhere between one in twelve and one in twenty people in Europe and North

America.[28] As with allergic disorders, a predisposition appears to run in families. One in three people with lupus, for example, also suffers one or more other types of autoimmunity, and half report having another family member with an autoimmune disease.[29]

Clearly, whatever was driving the Western world's epidemic of Th2-type allergies and asthma was not doing so at the expense of Th1-type autoimmunity. Rather, something appeared to have loosed the brakes on both immune locomotives, and the consequences hinged on a person's genetic predispositions. But even as the concepts of "protective" infection and Th1-Th2 "imbalance" fell by the wayside, the hygiene hypothesis took a new turn with growing evidence that the germs that make us sick may in fact be of less importance than the harmless multitude that have been engaging our immune systems for hundreds of millions of years—that is, long before the advent of civilization made contagious disease a part of everyday life.

CHILDREN IN THE COWSHED

Erika von Mutius strides down the long, uncluttered hallway that leads to her asthmology clinic at the University Children's Hospital of Munich, the click of her low heels bouncing off the gleaming linoleum and bright yellow metal cabinets that reach from floor to ceiling. Inside her examining room, flowery drapes and a windowsill jade plant help soften the stark environment, as does von Mutius herself, her short, dark, bouffant hairdo already tousled, her white lab coat enlivened with pins—a golden sun, a jumping child, a butterfly.

On the examining table, a chubby golden-haired grade-schooler sits stripped to his waist, the first of the parade of well-scrubbed young children who will fill her clinic day. "*Guten Morgen*," von Mutius says with a smile that engenders a shy grin from her patient. The boy's mother describes the nighttime cough and daytime shortness of breath that brought them here. Von Mutius asks him for deep breaths beneath the cold touch of her stethoscope. "*Gut, gut*," she encourages him, listening for the familiar internal wheeze.

When von Mutius began training in pediatrics at this same hospital in the mid-1980s, she quickly recognized the advantages of a subspecialty in allergy and asthma. The intertwined rates of the two disorders had already begun their dramatic upward trajectories, ensuring that she would have no shortage of young patients. On the research front as well, theories about the increase were as plentiful as solid epidemiology was scarce. "Working on the asthmology clinics, I saw the importance," she says. "I saw what a terrible disease this is not only for children but also for their anxious parents."

In 1989, as a young allergist in training, von Mutius began comparing the allergy and asthma rates of children in urbanized Munich with those living in the small towns of the surrounding countryside. At the time, most investigators were blaming air pollution for the strikingly higher rates of asthma among city kids, although any number of environmental or lifestyle differences might have contributed to the disparity. Von Mutius hoped to pin these factors down using family questionnaires and medical histories. But the results were disappointing. While allergy and asthma appeared to be slightly less prevalent outside the city, they were anything but rare. Stranger still was the one subset of country children who enjoyed a markedly lower rate of the disorders. "When my statistician ran his numbers," says von Mutius, "he told me, 'There's just one big signal here. Kids in homes with coal and wood heating seem to be protected.'" It made no sense at all. If anything, dirty-burning fuels should have exacerbated, not protected against, asthma. "Needless to say, we didn't publish," says von Mutius. "That was so contradictory to everything thought at the time."

A couple of months later, von Mutius forgot all about the rural-urban comparison. On November 9, 1989, the Berlin Wall fell. Reunification, von Mutius immediately realized, gave her the unprecedented opportunity to compare asthma and allergy prevalence in two ethnically identical populations living in starkly different environments. At the time, few nations surpassed West German standards of high air quality and emission controls. Meanwhile, East Germany choked beneath its legacy as the heavily polluted industrial center of the crumbling Eastern Bloc. "In Europe there was really no such place like Leipzig and Halle in terms of pollution," von Mutius recalls.

Over the next two years, she and her small team of nurses and physicians administered allergy and asthma tests to more than seventy-five hundred children on both sides of the former east-west divide, comparing their results against detailed medical histories and parental questionnaires aimed at ferreting out signs of allergy—from eczema to hay fever—and differences in lifestyle. This time the statistician had even more confounding results. That air pollution was taking a toll on East German children seemed clear from their higher rates of bronchitis—an indicator of airway injury akin to that seen in smokers. But according to the statistician's analysis, the same kids were three times *less* likely to suffer from hay fever and a third less likely to have full-blown asthma than were their West German peers. Further analysis correlated the asthma to higher rates of allergies to respiratory allergens such as mites, animal dander, and pollen. Well over a third of the West German children proved allergic, compared with fewer than one in five East Germans.

When published in *The American Journal of Respiratory and Critical Care Medicine* in 1994, von Mutius's report grabbed attention. Many dismissed its results as improbable. "I was just a tiny little doctor at the time," she says. But others saw a possible parallel with Strachan's newly popular hygiene hypothesis, with its emphasis on the protective role of exposure to germy children. "In this former socialist country, almost all East German mothers were working, with their children starting day care as babies," von Mutius explains. And children attending day care were always picking up colds and sundry other infections from their classmates.

Von Mutius became intrigued with Strachan's idea that childhood infections might be the missing link. But when she went back to Leipzig five years after reunification, she found that while asthma remained rare in older grade-schoolers, hay fever rates had doubled and those of eczema had increased by half.[30] Having been born years before reunification, most of these children had spent their infancy in day care. What had changed since that time was their families' rapid adoption of a Western lifestyle, including a dramatic shift in diet from farmer's market staples such as unpasteurized dairy products and unwashed, fresh-picked produce to a reliance on imported, largely processed foods such as margarine and canned vegetables.

Meanwhile, a conversation with a colleague across the Alps made von Mutius recall the unusual results of her first study, with its strange finding of lower asthma rates among children living in homes heated with wood or coal. The Swiss epidemiologist Charlotte Braun-Fahrlander had conducted a similar study of allergy rates among more than fifteen hundred children living in rural villages. At the insistence of a local doctor, she had included a query about farming in her family questionnaire. The village physician claimed that the children of dairy and pig farmers in his community didn't get hay fever. His observation proved accurate. Braun-Fahrlander's study showed that farm children were three times less likely to develop allergies than were their classmates living in town. And the more farming their families did, the better: the children of full-time farmers were half as likely to have allergic sensitivities as were the children of part-timers.[31] "Suddenly it made sense," says von Mutius of her earlier findings, "because farming in much of Europe is the same thing as heating by coal and wood, because that's the traditional way in old farmhouses." What was it about farm life that conveyed protection?

Beginning in 1998, von Mutius and Braun-Fahrlander sent a small army of graduate students and nurses across the Bavarian and Swiss countryside to interview farm families, take blood samples, and vacuum dust from stables, kitchen floors, children's bedding, and household air. Early and frequent exposure to stables turned out to be the single lifestyle factor associated with the lowest rates of asthma and respiratory allergies, which remained less than one in one hundred among children who spent their first five years of life around livestock—ten times lower than the rate among city kids.[32] Analyzing their dust samples, the research team also found a direct inverse relationship between allergy and asthma and the level of lipopolysaccharide, a chemical marker of bacteria, in children's mattresses.[33] The highest levels of lipopolysaccharide correlated with the lowest rates of allergic disorders, including asthma.

Big families, day care centers, and stables—the emerging commonality appeared to be the kind of bacteria abundant in dirt, diapers, and dung. Still other studies found that the allergy-protective effect of early exposure to pets proved much stronger with dogs than with cats, inviting speculation that the crucial difference might reside in where and what a pet sniffs and licks.[34] Adding to this suggestion of a "fecal factor," in 2000 Italian epidemiologist Paolo Matricardi published the results of his analysis of the antibodies found in the blood of 1,659 Italian air force recruits. By their nature, antibodies reflect a person's previous exposure to both respiratory bugs (such as those for measles, mumps, and rubella) and "orofecal" germs, or those spread by water and hands invisibly contaminated with the intestinal bacteria of an infected person. In the latter category Matricardi looked for antibodies to the stomach bug *Helicobacter pylori*, the intestinal parasite *Toxoplasma gondii*, and the hepatitis A virus. Few of the recruits would have actually gotten sick from exposure to the latter three germs, which typically infect without causing symptoms. But Matricardi found that allergies and asthma afflicted fewer than 7 percent of the cadets who had been exposed to two or three of these germs. By contrast, allergic disorders affected over 20 percent of those who had never met any of the bugs—a threefold difference.[35]

The hygiene hypothesis that emerged at the end of the twentieth century looked quite different from that first proposed by Strachan in the 1980s. According to the hypothesis in its later form, what protected against immune dysregulation was not sickness but early and continual exposure to microbes, especially the kind of bacteria that children regularly inhale and swallow when they are around other children, animals, and unsanitized water. Or more accurately perhaps, the lack of this exposure appeared to promote immune disorders in people with an underlying predisposition. Still, the world's immunologists understood that they had only begun to glimpse the regulatory mechanisms that normally keep the immune system in check. The idea that exposure to bacteria could provoke anything but a disease-fighting inflammation had yet to make it into textbooks and remained somewhat controversial, despite a growing literature of published research.

TEACHING TOLERANCE

In 1989, when Strachan published his hygiene hypothesis and von Mutius was discovering the allergy-protective features of farm life, another breakthrough was in the making. Newly appointed Stanford professor Dale Umetsu, an immunologist fresh out of his postdoctoral work at Harvard, had just published his proof that Th2 cells operate in humans.[36] Over the next decade, as Umetsu combined laboratory research with patient care, he became increasingly intrigued by the idea that infection—or something that mimicked it—might skew the immune system away from a Th2 cell response and so protect against the debilitating asthma attacks that plagued his patients at Stanford's Lucile Packard Children's Hospital.

Working in his laboratory with allergy-prone mice, he devised a vaccine made of the killed cells of the gastrointestinal bug *Listeria monocytogenes*. In modern times this bacterium has become infamous for causing food poisoning when it overgrows in cured meats such as hot dogs. But for most of human history—right up to the last quarter of the twentieth century—virtually everyone encountered tiny amounts of it in their drinking water, on fresh produce, and in the soil and dust around their livestock. Studies showed that listeria could induce a strong nonallergic, or Th1-type, immune response, and Umetsu speculated that exposure to it might help turn an allergy-prone immune system away from the opposing Th2-type response. In 1998 his research team made an experimental allergy "vaccine" by combining killed listeria cells with an offending allergen, in this case shellfish protein. They found that a single injection completely protected lab mice that had been so severely allergic to shellfish that ordinarily a trace amount would have triggered deadly anaphylaxis.[37]

In line with the thinking at the time, Umetsu and his colleagues thought that listeria's allergy-curing magic had to do with rebalancing the immune system's Th1-Th2 teeter-totter. But when they tried injecting Th1 cells into allergy- and asthma-prone mice, they were shocked to find that they worsened inflammation in the animals' lungs.[38] The unexpected results prompted Umetsu to look at the inflammation-calming effect of bacteria in a new way.

Just because the lungs of a nonallergic mouse—or person—show little inflammation, Umetsu reasoned, that doesn't mean that its immune system doesn't recognize and respond to allergens. Blood work revealed that everyone carries antibodies to almost everything in their environment. Rather than ignoring the antigen, Umetsu realized, the immune system of a nonallergic person recognizes it as nonthreatening and actively produces a tolerance response that *reduces* inflammation. The source of allergies therefore was not an imbalance in Th1 and Th2 immune cells—each of which produced a type of inflammation—but a lack of tolerance.

With this in mind, in 2004 Umetsu took a closer look at the Th1 cells that his listeria-and-allergen vaccine had induced in laboratory mice. These were no ordinary helper T cells. They didn't produce the kinds of signaling molecules that marshaled the immune system to attack infected or otherwise damaged body cells. These Th1 cells produced interleukin-10—the immune system's universal signal to "stand down."[39] In this respect they resembled another recently discovered breed of immune cells, which had been dubbed regulatory T cells. T-reg cells helped keep the immune system from attacking the body's own tissues—that is, from triggering autoimmunity.

Japanese immunologists had discovered a general class of regulatory T cell the previous year.[40] Moreover, they had shown that crippling the T-reg cell's master control molecule (Foxp3) resulted in a fatal combination of severe allergies, inflammatory bowel disease, and autoimmune disorders.[41] Umetsu's lab went on to use their listeria-allergen vaccine in lab mice and dogs to induce tolerance to allergens such as eggs, nuts, and mold and so reverse life-threatening respiratory and food allergies.[42] Significantly, they helped show how harmless bacteria such as killed listeria cells could produce this tolerance. And they did so by engaging a less well understood but far more ancient branch of the immune system.

INNATE IMMUNITY

T cells, with their ability to recognize and target specific antigens, are part of a branch of the immune system called *adaptive*. The adaptive immune system works with stunning efficiency: the first time it encounters a specific antigen, it produces long-lived memory cells that continue to circulate throughout the body. The second time the antigen shows its molecular face, these memory cells rapidly proliferate to give rise to a newly cloned army. As a result, vaccines and some kinds of infections produce lasting, even lifelong immunity.

The mechanisms of adaptive immunity explain how the immune system responds to already encountered threats. But for most of the twentieth century, immunologists had little understanding of how a naïve, or untrained, T cell distinguished between antigens deserving of attack (those associated with infectious microbes or diseased cells) and those that should be tolerated (those associated with food, the dreck and dander of everyday life, and the beneficial bacteria of the human colon).

Then, in the late 1980s and early 1990s, researchers discovered an ancient family of proteins, dubbed toll-like receptors,[43] on the surface of immune "scout" cells whose job it was to capture and present foreign matter to the immune system's major generals, the T cells. As scientists studied these proteins further, they realized that toll-like receptors were pattern-recognition molecules that responded to generic markers unique to microbes. For example, the first two toll-like receptors discovered, TLR1 and TLR2, bind to lipopeptides (fatty proteins) on the surface of gram-positive bacteria. Viral RNA activates TLR3. TLR4 registers the distinctive rub of a gram-negative bacterium's lipopolysaccharide coat. TLR5 responds to the chemical tickle of a swimming bacterium's whiplike tail, or flagellum. Still other toll-like receptors bind to the "naked genes," or unmethylated DNA, used by all bacteria and many viruses.

In plants and invertebrate animals, toll-like receptors stud the surface of many kinds of cells. With the evolution of adaptive immunity in jawed fishes and higher animals some 400 million years ago, cells with toll-like receptors took on the specialized role of capturing microbes and foreign substances and presenting their predigested bits and pieces as "antigen" to T cells.[44] Chief among these antigen-presenting cells are dendritic cells—octopus-like cells that concentrate beneath the mucous membranes of the respiratory and digestive tracts, insinuating their long arms up through the organs' delicate lining to sample whatever bacteria, food, and other substances might be wafting or bumping by.[45]

With the discovery of toll-like receptors, immunologists came to view dendritic cells as the immune system's advance scouts and military advisers. Research showed that these cells did more than simply present antigens to T cells for recognition; they did so along with a burst of instructive cytokines, which allowed them to report whether a particular antigen was worthy of attack.[46]

Then, in 2001, immunologists at Rockefeller University announced that they could make dendritic cells do just the opposite—turn *off* an adaptive immune response. They did so by repeatedly stimulating the cells' toll-like receptors over a prolonged period.[47] Other research confirmed this counterintuitive finding, showing that in the absence of a clear danger sign (such as tissue damage) dendritic cells respond to the *continual* presence of bacterial products such as lipopolysaccharide by secreting that calming cytokine of tolerance, interleukin-10.[48] This in turn prompts a naïve T cell to mature, not into an inflammation-driving Th1 or Th2 cell but into a regulatory T cell that broadcasts more immune-calming cytokines.[49]

When Dale Umetsu and his Stanford colleagues used their killed-listeria vaccine to cure mice of asthma, they found they could take dendritic cells from the vaccine-treated mice, inject these cells into other asthmatic mice, and cure their asthma as well.[50] They also discovered they could alleviate mouse asthma with injections of nothing more than snippets of bacterial DNA.[51] Umetsu says that this demonstrates that bacteria or their DNA can turn off the inflammation behind asthma, not by simply tipping the balance from an allergic Th2 immune response to cell-killing Th1 response, but by engaging a newly discovered pathway of tolerance.

Such work dovetails nicely with the farmhouse studies of von Mutius and their implication that early and constant exposure to bacteria in bedding and household air helps protect children from allergy and asthma.[52] As further evidence, studies on farm families have shown that the relatively few farm children who develop allergies and asthma often have mutations in the genes for the toll-like receptors that register the presence of bacteria.[53]

In 2005 Umetsu returned to Harvard, where he continues to use killed bacteria to plumb the mysteries of the immune system and perhaps develop a cure for allergies and asthma. Meanwhile, he is intently following research in Great Britain that has produced what may turn out to be an even stranger type of bacterial vaccine than his killed listeria, and one with even more far-ranging benefits.

THE DIRT VACCINE

John and Cynthia Stanford have returned to Mill House, their eighteenth-century farmhouse in Tonbridge, Kent, for a brief respite between trips abroad. Their parade of patients, on a late-winter morning in 2006, begins before morning tea. John Stanford, recently retired as head of microbiology at the University College of London Medical School, dresses in flannel and heavy trousers, his full white beard and tousled hair adding to the impression of jovial eccentricity. Standing in the cozy warmth of an oven-heated kitchen, he bends to receive a kiss on the cheek from the first arrival, a plump, rosy-faced English grandmother in brown cardigan and herringbone skirt. Sue Hamilton-Miller has been coming for her bacterial injections for over two years, following a diagnosis of inoperable melanoma that had spread to her lungs. "They were wheeling me into surgery when the doctor came up to me with the results of the X-rays," she confides. "'I'm so sorry,' he told me. 'There's no sense in operating.'"

Cynthia Stanford shows Sue into a sitting room layered in a tapestry of Persian rugs that extends across the walls as well as the floors. Cynthia is as small and slender as her husband is broad-shouldered and tall. Both share a milky English complexion heavily sun-spotted from decades spent trekking across much of the Third World. When John ducks through the doorway to join them, he holds a syringe filled with heat-killed cells of *Mycobacterium vaccae* dissolved in a borate solution that helps break apart the microbe into its constituent parts. Sue pulls off her cardigan and pushes up her sleeve to uncover a dozen small pink bumps across her upper arm. They are the fading welts from past injections. "They flare with each new jab," she says, "then settle back down again."

Sue found her way to the Stanfords after her oncologist agreed that conventional treatment offered little hope, given the advanced stage of her cancer. Her husband, Jeremy, a medical microbiologist with London's Royal Free Hospital, had been following the Stanfords' immunology work for years, and John and Cynthia gladly added Sue to their roster of patients receiving their experimental vaccine on a compassionate-use basis.

Within weeks of Sue's first *M. vaccae* injection, tests showed that her tumors had stopped growing. With weekly, then monthly, injections, they then regressed without either chemotherapy or radiation. At the end of the year, her chest X-rays showed that her lung tumors had shrunk to half their starting size. She continues to come to the Stanfords for follow-up injections of what the British media has dubbed the "Dirt Vaccine." The roster of ills that it's purported to relieve tempts comparison to snake oil: not just cancer but also a range of allergic disorders, including asthma; several forms of autoimmune disease; even tuberculosis and leprosy. "All these diseases result from an immune system out of balance," John asserts, "from an ineffectual or damaging immune response instead of one focused on clearing the body of diseased cells."

The story of Mycobacterium vaccae's development into a form of immunotherapy began in the early 1970s. John and Cynthia, with their five young children in tow, were crisscrossing Uganda, taking soil samples. They were searching for something in the soil that might explain why the Bacillus Calmette-Guérin (BCG) vaccine so effectively protected Ugandan children from both tuberculosis and leprosy. Mycobacterium tuberculosis and Mycobacterium leprae, the microbes causing these two diseases, rank among a half dozen troublemakers in a large family of soil bacteria that are distinguished by their waxy, water-repelling cell walls. BCG, the world's most widely used vaccine, contains a weakened strain of yet another mycobacterium, M. bovis, which can cause tuberculosis in cattle. When it works, BCG prompts cross-immunity against human tuberculosis, in the same way that Edward Jenner's historic cowpox vaccine protected against smallpox.

But the BCG vaccine has long frustrated the medical world with the wide variability of its protection. It's more effective in infants than in older children and adults, and its overall protection rate ranges from as high as 80 percent in some countries to as low as zero in others.[54] Even within Uganda, a nation approximately the size of Minnesota, the vaccine's effectiveness varies widely from region to region, reaching a peak around the isolated shores of Lake Kyoga, a shallow inland sea where nomadic tribes bathe and draw cooking water.

Tuberculosis researchers such as the Stanfords' colleague John Grange had long proposed that BCG's variability might stem, in part, from a kind of natural booster effect conveyed by cross-resistance to harmless mycobacteria in a person's environment.[55] Abundant in soggy, oxygen-rich swamps and shores, most mycobacteria make a living, not by infecting people, but by breaking down and recycling dead vegetation. With this in mind, the Stanfords began trawling the country's soil and waters in a crisscross fashion to see if they could find one or more kinds of mycobacteria that strongly correlated with a good BCG response among local children. From Lake Kyoga's muddy shoreline, John isolated M. *vaccae*. Of all the mycobacteria in the Kyoga soil, he explains, M. *vaccae* had the kind of complex, antigen-rich cell wall that he believed would provoke the strongest immune response.

Initially, John's work with M. *vaccae* centered on creating a more effective vaccine against tuberculosis and leprosy, either alone or in combination with BCG. Both avenues produced modestly good results.[56] More ambitiously, John hoped that M. *vaccae* might help those in the early stages of these diseases to clear their infections. Both M. *tuberculosis* and M. *leprae* cause chronic and worsening disease in just 10 percent of those infected. In the other 90 percent, a strong Th1-type immune response clears the body of infected tissue. The progressive lung destruction of chronic tuberculosis and the extensive bone, cartilage, and nerve damage of leprosy result when the immune system opts instead for a mixed Th1 and Th2 response that razes tissue surrounding the infected cells, continually walling off the infection rather than eliminating it. If M. *vaccae* engaged the immune system as strongly as John hoped, a shot of the killed bacterium might be one way to redirect a patient's immune system toward the desired curative response. As such, it might offer hope in cases where antibiotics fail to eradicate an extremely drug-resistant strain.

In 1975, before using their experimental M. *vaccae* vaccine on anyone else, the Stanfords injected each other. "If we weren't willing to use it on ourselves, we could hardly present it to other people as harmless," Cynthia explains. In retrospect, she may have been the first to benefit from its effects. "After the little pink bump at my injection site went down, I forgot about it," she recalls, "until we were in the middle of another cold, wet English winter, and I realized that I hadn't suffered from my usual Raynaud's."

Raynaud's syndrome, an autoimmune disorder that shuts down circulation in a person's extremities, can produce painful numbness of fingers and toes, typically in response to cold or stress. It ran strongly through Cynthia's family, and so her mother, sister, and youngest daughter, Thomasina, lined up for injections every few months through the following year. Their symptoms likewise improved. "Then we noticed that other things were disappearing too," she says. Her mother's spinal cancer regressed, her sister's arthritis disappeared, and her daughter's asthma took a two-month hiatus, never to fully return. Doctors field-testing *M. vaccae* at leprosy and TB clinics in Asia likewise noticed unexpected benefits. From India came back reports of patients whose scaly psoriasis resolved following vaccination.[57] An autoimmune disorder, psoriasis results when the immune system kills off healthy skin tissue and the dead cells pile up to form thick, silvery scales.

M. vaccae's beneficial side effects had precedent. As far back as the 1960s, doctors had reported that children who received the BCG vaccination enjoyed slightly lower rates of allergy and asthma, autoimmune diseases such as type 1 diabetes, and even leukemia.[58] Since the late 1970s, the BCG vaccine had even been used as a cancer treatment, most successfully against the small tumors of superficial bladder cancer.[59] Along these lines, in the early twentieth century the American bone surgeon William Coley pioneered the use of bacterial vaccines for treating sarcomas—malignant tumors of bone, muscle, and other connective tissue. Coley's vaccine consisted of a mixture of killed *Streptococcus pyogenes* and *Serratia marcescens*.[60] The American Medical Association endorsed the vaccine in 1936, but it rapidly fell into disuse with the rising popularity of radiation and chemotherapy the following decade.[61] More enduring has been the intriguing idea that exposure to certain bacteria, be they alive or dead, might somehow jolt the immune system into effective action. *M. vaccae*, the Stanfords have come to believe, is one of many kinds of bacteria in the natural environment upon which the immune system has come to depend for optimum performance. "We've identified a whole series of related bacteria that engage the immune system in such a beneficial way," John says.

Soon after the Stanfords brought *M. vaccae* back from Uganda in 1975, John began working with his University College colleague Graham Rook, an immunologist fascinated with the underlying mechanisms through which bacteria interact with the immune system. Their early studies with laboratory animals and several small clinical trials confirmed that an injection of dead *M. vaccae* cells could both relieve allergies and asthma and enhance the immune system's cancer-fighting abilities.[62] In 1992, with the backing of University College London, Stanford and Rook formed SR Pharma, a publicly traded company through which investors could fund the gauntlet of clinical trials needed to turn *M. vaccae* into a bona fide medical treatment. The encouraging results of their early work and the medical detective story behind

M. vaccae generated television documentaries and magazine features about the so-called Dirt Vaccine, with headlines such as "Bring on the Germs" and "Let Them Eat Dirt!"[63]

Then came the disappointing results of a media-hyped study with advanced lung cancer patients who were also receiving conventional chemotherapy. The study found no significant increase in survival time, only a markedly higher "quality of life" among patients who received *M. vaccae*. The group reported being far more active, comfortable, alert, energetic, and sociable than did those who received a dummy vaccine.[64]

The independent research team hired to run the trial bollixed the recruitment, the Stanfords say, by mixing together patients with different kinds of lung cancer. A statistical reanalysis of the study showed that *M. vaccae* did in fact increase survival by an average of 135 days for patients with adenocarcinoma but had no survival benefit for squamous cell lung cancer.[65] Regardless, publication of the initial results in 2001 sent SR Pharma's stock crashing.[66] The company's business managers moved quickly to cut losses. Within a year they had suspended clinical trials with *M. vaccae*, removed Rook and Stanford from the company's primary board of directors (they remain advisers and shareholders), and changed the company's focus to acquiring more futuristic biotechnology.

Rook barely looked up from his laboratory bench. The Stanfords, who were heading into retirement, formed a new privately held company, BioEos, in 2004. With SR Pharma holding the patent to *M. vaccae*, they began developing a half dozen kinds of closely related bacteria for use as veterinary products. John injected one of these bacterial-cell preparations into the equine sarcoid tumors of his sister's five-year-old thoroughbred. After fourteen injections, thirteen of the horse's fourteen tumors disappeared. Photos show that the largest tumor, around three inches in diameter, shrank to a flat scar a quarter of an inch wide. Two years after the last injection, none of the tumors had returned. In 2006 the Stanfords began using the same bacterium in a larger trial of several hundred horses suffering from "sweet itch," a severe midge-bite allergy that plagues millions of horses each spring. The allergy disappeared entirely in a quarter of the animals and eased off significantly in around half.[67]

Preliminary studies show that several of the Stanfords' other bacterial preparations may prove more profitable. Some speed the growth of piglets and calves as well as farm-raised fish and prawns.[68] John speculates that these immune-priming bacteria help the animals clear low-grade infections that would otherwise sap their energy. If larger trials confirm the benefit, the products could become an alternative to growth-promoting antibiotics, which were recently banned by the European Union. "Then we could use this money to fund clinical trials for treating allergies and asthma and other conditions in humans," says John. "That's the goal. This is the means."

Meanwhile, cancer patients continue to find their way to the Stanfords' home in the south of England. Following Sue Hamilton-Miller, the recovering melanoma patient, visitors on a recent late-winter day included long-term survivors of colon, rectal, and pancreatic cancer—one of whom had opted for M. vaccae in lieu of a colostomy and two of whom had been given terminal diagnoses years ago. All claimed they knew when they needed another injection because a certain sparkle left their lives. "It's hard to explain," says Hamilton-Miller, "but after my jab, I'll wash all the windows and then look around wondering what else I can do."

When asked about those who have died, the Stanfords bemoan a common pattern. "Once they start getting better, their doctors tell them, 'Well, now you're strong enough for chemotherapy,'" says Cynthia. "And the next thing we know, they're dead." That's not so surprising, says John, given that chemotherapy is infamous for suppressing the immune system. Still, the Stanfords appreciate that their results can make M. vaccae sound like a cross between laetrile and happy pills. "That's the problem, isn't it?" says Cynthia. "It sounds like we're steaming crazy."

OLD FRIENDS

Back at University College, in an office overlooking London's scholarly Bloomsbury district, the Stanfords' longtime collaborator, Graham Rook, remains equally obsessed with M. vaccae, though perhaps with less interest in its current medical use than in its intriguing effects on the immune system. A slightly built, ramrod-straight Englishman, Rook wears the Cambridge man's uniform of pale blue shirt, khaki dress pants, and crewneck sweater. Charts of immune-signaling chemicals and cell-surface markers cover the walls of an otherwise undecorated room piled with books and papers. Unlike the Stanfords, Rook continues to work as an active consultant with SR Pharma. At the same time, his academic position provides him with a laboratory full of postdocs and graduate students eager to pursue research on M. vaccae.

Like Umetsu's team in the States, Rook's has been injecting killed bacteria, in his case M. vaccae, into allergic and asthma-prone mice and correlating changes in their symptoms with underlying shifts in different types of immune cells. Specifically, Rook has found that a shot or oral dose of M. vaccae greatly reduces allergic and asthmatic reactions in mice while stimulating the proliferation of regulatory T cells and dendritic cells.[69]

He and his colleagues have already sliced and diced M. vaccae into its chemical parts to isolate what exactly it is about this bug that seems to goad an out-of-whack immune system to straighten up and fly right. The isolation of this key molecule, or active ingredient, began with Rook's observation that stripping the bacterium of its water-repelling envelope destroyed its allergy-protective effects. Next Rook sent an extraction of cell-coat molecules to a chemist to split into fractions. "He'd split them into two pots by some kind of chemical criteria, and we'd test them both, sending back the one that protected the mice," says Rook. "Then he'd split it into two pots again. When we got to the point where he told us, 'There's hardly anything in here,' we asked him to find out what was." It was a single lipid molecule from the bacterium's waxy coat, which Rook's crew has since synthesized in the laboratory to produce the patented product SRP312 (for SR Pharma product 312)—a synthetic stand-in for the whole M. vaccae bacterium.

He and his lab colleagues now work with the whole organism and with SRP312. Just two micrograms of SRP312—two one-millionths of a gram—protects egg-allergic mice from the kind of overwhelming asthmatic attack normally produced by a squirt of egg protein. Delving deeper, Rook and his colleagues have analyzed the lung secretions of SRP312-treated mice and found the tissues to be suffused with the anti-inflammatory cytokine interleukin-10—around eight times more than is found in the lungs of untreated mice and around four times as much as found after a shot of killed M. vaccae.[70]

At the same time, Rook has gone farther than anyone in reworking

Strachan's original hygiene hypothesis, with its largely discredited emphasis on sickness. "The whole idea of infection being good for you turns out to be rather silly when you think about it," he says. "Infection means inflammation, and that makes things worse." Rook distinguishes between disease-causing "infection" and harmless "colonization" by the body's normal microflora and environmental bacteria that pass through the body in food and water. He also points out that contagious diseases became a part of daily life only with the crowding of civilization some five thousand years ago. By contrast, the mycobacteria of untreated water and mud and the lactobacilli and other organisms abundant on fresh or crudely stored food have been with us throughout our evolution.

"I rather like the idea of 'old friends,' as in organisms that were with us all the time, every day, inescapable," Rook says. In nature's way of embracing the inescapable, the immune system responded to life's ocean of harmless bacteria with regulatory cells that secrete biochemical messages of tolerance, he says. "If you don't have that normal background level of bacteria, your body doesn't have a background level of regulatory cytokines, and your immune system just overresponds to everything."

Rook's "old friends" theory of immune regulation recognizes three major groups of immune-calming organisms. One is the bacteria abundant in the natural environment. The second is the body's own commensal, or symbiotic, bacteria, which antibiotics can disrupt and alter to a significant degree. More than a half dozen large studies, he points out, have confirmed that children who receive antibiotics in the first year of life have more than double the rate of allergies and asthma in later childhood, even if the prescribed antibiotics were to treat a non-respiratory problem.[71] Animal research by University of Michigan immunologists Mairi Noverr and Gary Huffnagle may explain this effect in part: they made mice highly susceptible to respiratory allergies and asthma by colonizing their guts with *Candida albicans*, a fungus or "yeast" native to the human intestinal tract, then disrupting the animals' intestinal bacteria with antibiotics. They concluded that antibiotics allow the intestinal candida to overgrow to harmful levels and that the resulting inflammation predisposes to allergies.[72]

Other changes in intestinal microflora, unrelated to antibiotic use, may likewise predispose a child to allergies. Babies born by cesarean section, for example, have double the rate of food allergies as vaginally delivered infants—a finding that suggests that the mother's vaginal and anal bacteria deliver benefits that are missed during a surgical birth.[73] Other research has revealed that allergic infants tend to have more adultlike intestinal flora (abundant in coliform and even staph bacteria) than do nonallergic infants (who have more lactobacilli and bifidobacteria).[74] And this type of shift to adult intestinal flora tends to be much more common among children in highly sanitized countries such as Sweden than in less developed countries such as neighboring Estonia, where children enjoy much lower rates of allergy and asthma.[75]

"Obviously, it's not just one thing," says Rook, "or we'd all be suffering from allergy and asthma." Rook counts helminths, or intestinal worms, as the third major category of "old friends." "It drives a lot of parasitologists mad that I call them that," he says. "But the body has to treat them as friends because, once you're infected, if you don't turn on the regulatory cells and stop the immune response, you destroy your lymphoid system and end up with elephantiasis." It appears that evolution has shaped the Th2 arm of the adaptive immune system to react strongly to early signs of helminth infection but then to accept the inevitability of a stubborn infestation by turning off the immune response.

As evidence, in 1999 the Dutch biologist Maria Yazdanbakhsh found that eliminating intestinal worms from infected children in African villages immediately predisposed them to allergies. Within a few months of being treated, the children tested allergic to twice as many substances as did their still-parasitized peers.[76] No one is proposing that we leave children infected with parasitic worms, Rook emphasizes, for a heavy infestation can stunt growth and cause dysentery, anemia, and even mental retardation. Rather, such discoveries are prompting researchers to try to replicate the immune-calming benefits of intestinal parasites without also replicating their harm. University of Iowa gastroenterologist Joel Weinstock, for example, has used a drink spiked with the eggs of a harmless pig whipworm to relieve severe inflammatory bowel disease in patients, with an 80 percent success rate and no ill effects.[77] Other scientists have shown that they can protect allergy-prone lab animals by deliberately infesting them with similar parasites.[78]

BEYOND IMMUNITY

Most recently Rook has become intrigued by signs that our reduced exposure to environmental bacteria and other "old friends" may be affecting our quality of life in ways that extend beyond the immune system. The new interest stems from a collaboration between Rook and neuroscientist Stafford Lightman, of the University of Bristol. Rook and Lightman had been studying the hormonal changes brought about by different types of immune responses. In their experiments, they would induce a Th1 type of infection-clearing immune response in lab mice by injecting the animals with *M. vaccae*. To induce a Th2 response, they injected allergy-prone mice with egg protein. By chance, in an adjoining laboratory, a young neuroscientist was studying brain cells that produce the hormone serotonin. This young researcher, Christopher Lowry, already knew that activating serotonin-producing cells in one part of the brain improved mood and activating those in another increased alertness and agitation.[79] This discovery went far in explaining why the class of antidepressants known as selective serotonin reuptake inhibitors (Prozac, Zoloft, and others) can have the unwanted side effects of triggering insomnia and anxiety. The perfect drug, Lowry surmised, would turn on only the serotonin-producing brain cells that enhanced a person's mood *without* activating those that result in anxiety and hyperalertness.

All this had nothing to do with Rook and Lightman's research, or so they thought. "Chris Lowry was so obsessed with looking at these two kinds of serotonergic neurons," recalls Rook, "that he came in saying, 'Oh, you've got these lovely mouse brains. Can I look at them under my microscope?'" And so he did, after staining the brain tissue to highlight cells actively exchanging serotonin messages. Says Rook, "He came rushing back into Lightman's lab, shouting, 'You've got to look at this! They're flashing at me, and all you've done is inject your mice with these silly dead bugs.'" The brains of the *M. vaccae*–treated mice showed serotonin activity only in the brain-stem cells associated with elevated mood. "Prozac without the side effects," says Rook.

The curious finding began making sense when Rook uncovered research showing that injections of interleukin-10, the cytokine most associated with regulatory immune cells, counteracted lethargy and social withdrawal in ailing lab rats.[80] Next he learned of research by University of Colorado psychiatrist Marianne Wamboldt, who discovered a common genetic predisposition to both depression and allergic disorders. While it could be argued that the stress of chronic allergies may lead to psychological problems, Wamboldt's studies with twins showed that if one twin had an allergic disorder such as hay fever, eczema, or asthma, the other twin was more likely to suffer depression, even without allergies.[81] The association proved to be far stronger between identical twins than between fraternal twins, another clear sign of a common genetic factor.

Could interleukin-10's "all is well" message extend beyond the immune system to the brain? As further evidence, Rook references a half dozen published studies showing that antidepressant drugs such as lithium boost levels of IL-10 and decrease allergy-associated cytokines such as interferon gamma.[82] Could *M. vaccae*'s IL-10 boosting powers explain that certain joie de vivre reported by Stanford's personal roster of cancer patients? Rook demurs. "I don't know about that. What he does almost seems to get into a placebo effect, an almost faith-healing kind of a thing." Still, says Rook, it explains the markedly higher quality-of-life scores in *M. vaccae*'s clinical trials with late-stage lung cancer patients.

"It would be interesting," he says, "if instead of blaming rising rates of anxiety and depression solely on the increased stress of modern living, we could be witnessing the effects of something as mundane as fewer bacteria in our food, water, and daily lives." He laughs when recalling the reaction he provoked from a group of psychoanalysts with just that suggestion. "This was after we'd had a few drinks together," he admits, "and I've got to say they got quite hostile, really quite rude." After several minutes of being shouted down as ridiculous for proposing that a cytokine—a mere molecule—could explain historic changes in people's behavior, Rook says he came up with the perfect riposte. "I said, 'Look here, there's a molecule called ethanol that you've just taken, and it's certainly changed your behavior.'"

PART IV

BUGS ON DRUGS

It's very clear that bacteria have been here much longer than we have, and as far as they're concerned, we may be just a passing feature in their history.

—STUART LEVY, M.D., 2000

A KILLER IN THE NURSERY

NEW GERM STRAIN
TAKES HEAVY TOLL

U.S. Studies Virulent Form of
Staphylococcus That Resists Antibiotics

SPECIAL TO THE NEW YORK TIMES
WASHINGTON—The Public Health Service said today . . .

Heinz Eichenwald didn't have to look past the headlines to know the
details of the story bisecting the metro page of his Saturday paper on
March 22, 1958. He was on a first-name basis with the damn germ, a
particularly nasty and drug-resistant strain of that age-old troublemaker
Staphylococcus aureus. As head of pediatric infectious disease at New
York Hospital, the square-jawed, intense young doctor had been following its deadly spread since the strain cropped up in California maternity wards and hospital nurseries in 1955. He also understood what had
propelled the issue into the headlines. Earlier that week, under intense
media pressure, administrators at Jefferson Davis Hospital, in Houston,
had held a press conference in which they admitted to the recent
staph-related deaths of sixteen newborns. Further, they reported that
over half their doctors and nurses were silently carrying the dangerous
strain. The drug-resistant bug, known in the medical community as
Staph. aureus phage type 80/81, shrugged off not only penicillin but
every antibiotic on the pharmacist's shelf—including sulfa drugs, tetracycline, chloramphenicol, erythromycin, and streptomycin.

Now the U.S. Surgeon General was announcing a national crisis. All this Eichenwald knew, and more. What made him almost spit out his coffee was the conclusion of the *New York Times* report:

NO OUTBREAK HERE

New York health and hospital officials said yesterday that there was no indication here of any outbreak of the infection. Dr. Morris A. Jacobs, Commissioner of Hospitals, said last night that he had confidence in the sterilizing and autoclave (sterilization under pressure) procedures used in hospitals here. Sterilization is especially important as a safeguard against the staphylococcus strain because it resists antibiotics.[1]

Well, at least Morris got the last part right, thought Eichenwald, who knew full well that New York Hospital had one hell of an outbreak on its hands, not that its administrators would have volunteered that information and risked scaring away patients. What hospital would?

Eichenwald had spent the previous year working twelve-hour days, hustling between the hospital's nursery and its pediatric ward—the former discharging healthy newborns and the latter filling with slightly older babies. That was when he began to suspect what was happening. Some of the returning infants arrived blue for lack of oxygen, their chests heaving with the strain of pulling air into infected, fluid-filled lungs. Others, red-faced and burning with fever, lapsed into unconsciousness as the staph spread to vital organs.

At the same time, Eichenwald began fielding phone calls from family physicians, many of them frantic in their inability to stop their patients' horrific impetigo, a type of staph skin infection. In many families the nasty rashes were spreading from newborns to parents and siblings, progressing to raw ulcers and ugly yellow boils. Some nursing mothers developed milk-duct abscesses so large that their breasts collapsed when the infection finally cleared. Old-fashioned scouring and draining of the wounds didn't clear the infections; nor did any of the sulfa drugs or antibiotics that had banished staph so easily just a few years before.

By the time the Houston story broke, Eichenwald was tracking the infections back to his own hospital. "We had a silent epidemic," he says, recalling his dismay a half century later. "It was silent because the babies looked perfectly fine when they left our newborn nursery to go home." Eichenwald helped coordinate the drastic countermeasures aimed at stopping the escalating outbreak. The hospital's orderlies and nurses doused floors and work surfaces with caustic chemicals. Workmen installed microbe-killing UV lights. Administrators hectored physicians into using more elaborate glove-and-gowning procedures. But nothing seemed to slow the strain's deadly spread. In desperation, the hospital tried putting newborns on antibiotics within an hour of birth. "If anything, they got infected faster," Eichenwald recalls. Administrators ordered the testing of all staff for the presence of 80/81, ordering home anyone harboring it. "A hospital keeping doctors and nurses at home on full pay? That tells you how bad this was," Eichenwald adds.

Still the epidemic worsened. Eichenwald began working with a nurse epidemiologist whose sole duty was to follow up on their discharged newborns. She would make a friendly phone call, nothing alarming: "Just checking on the baby . . . Feeding well? . . . Looking good? Any concerns?" Most of the discharged babies remained healthy. But others developed severe impetigo, eye infections, or pneumonia. Parents also reported how the infections were bouncing between family members for months and even years. One family actually abandoned its home. Another set of parents divorced under the strain.

At the same time, Eichenwald and his colleagues understood that almost everyone silently harbors one strain of *Staph. aureus* or another from time to time. Around one in four of us provides staph with a permanent home.[2] For the most part, the bug remains confined to a sweet spot of moist mucous membrane inside the nostrils and causes no problem. But *Staph. aureus* has always been an opportunist of the worst sort, more apt than any of our other resident microbes to invade and overgrow when given a chance. This was the microorganism that caused the boils that Job scraped from his body with broken pottery. And it was a creamy yellow colony of *Staph. aureus* that grew in the petri dishes of Alexander Fleming when, in 1928, he made his accidental and now-famous discovery of the bacteria-killing powers of *Penicillium* mold.

The virulent and drug-resistant 80/81 strain, bred in hospitals of the early 1950s, had a particularly nasty tendency to take advantage of the slightest break in the body's normal defenses. The baby pimples and diaper rashes of infancy became open doorways to the bloodstream. From there the bacterium could lodge in other organs, including the meninges, or lining of the brain. Just as potentially dangerous, the congestion of a baby's first cold gave 80/81 the opportunity to overgrow in the lungs and produce unusually severe and often rapidly deadly pneumonia.

While antibiotics didn't produce 80/81's virulence, they helped spur its spread. The introduction of these miracle drugs into hospitals at the end of World War II unleashed an unprecedented evolutionary force in the microbial world by killing off all drug-susceptible competition. In this way antibiotics fueled the success of any strain or species that was genetically endowed to resist their effects. In shrugging off every weapon in the modern antibiotic arsenal, the world's first pandemic superbug—*Staphylococcus aureus* phage type 80/81—forced infectious-disease specialists the world over to explore nonantibiotic solutions.

At New York Hospital, Eichenwald and infectious disease specialist Henry Shinefield conceived and developed a controversial program that entailed deliberately inoculating a newborn's nostrils and umbilical stump with a comparatively harmless strain of staph before 80/81 could move in. Shinefield had found the protective strain—dubbed 502A—in the nostrils of a New York Hospital baby nurse. Like a benign Typhoid Mary, Nurse Lasky had been spreading her staph to many of the newborns in her care. Her babies remained remarkably healthy, while those under the care of other nurses were falling ill. After helping to put a stop to the New York Hospital epidemic, Shinefield and several colleagues began traveling to other hospitals across the country and using 502A to help them end their outbreaks. Eichenwald remained behind at New York Hospital, sending out vials of 502A to hospitals around the world.

Today few people under age fifty appreciate what havoc virulent, drug-resistant staph 80/81 wreaked on the world, says Eichenwald, who in 2005 retired as the head of pediatric medicine at University of Texas Southwestern Medical Center. "We're talking about some of the few years in this century when infant mortality actually increased in the United States."[3] But the use of 502A never failed to stop a hospital outbreak, Eichenberg recalls. "I can't claim and I know Henry doesn't claim that 502A is what made 80/81 go away." But neither does Eichenwald credit the antibiotic that many hailed as the miracle that banished 80/81 and that would break the back of drug resistance altogether.

AN END TO BACTERIAL DISEASE?

Beginning in 1960, the British pharmaceutical giant Beecham and Bristol introduced methicillin, first in Europe and later in the United States. The much-anticipated drug had already made headlines in 1959, when doctors used an experimental batch to save the life of actress Elizabeth Taylor, who had fallen ill with 80/81 staph pneumonia while filming the blockbuster *Cleopatra*. To create methicillin, the first semisynthetic antibiotic, Beecham-Bristol chemists had tinkered with penicillin's structure, adding a chemical thorn to its beta-lactam ring. This foiled the deactivating enzymes produced by penicillin-resistant microbes. Methicillin, an injectible drug used primarily in hospitals, gave rise to several stomach-stable oral versions such as oxacillin, which came into widespread use outside of hospitals.

But the methicillin group of antibiotics had been in use for less than a year when doctors began encountering strains impervious to it.[4] And while the drug appeared to be effective against 80/81, in truth this virulent strain was already disappearing when methicillin was first introduced. "Not that these bugs ever really go away," comments Eichenwald. "They just disappear for a while and come back again in more resistant form."[5] By 1964 European hospitals were reporting large outbreaks of methicillin-resistant *Staph. aureus*, or MRSA, strains of which began leapfrogging into medical centers on other continents. For many medical microbiologists, the bacterial kingdom's rapid riposte to a laboratory-designed antibiotic ended their dream of a magic bullet that could banish bacterial disease. The new paradigm became that of an endless arms race with the microbial world. To quote the Red Queen of *Through the Looking-Glass*, "Here, you see, it takes all the running you can do, to keep in the same place."

The pharmaceutical industry rose to the Red Queen's challenge with impressive results. The 1960s and 1970s saw the debut of several entirely new classes of antibiotics that proved effective against a broad spectrum of bacteria on both sides of the kingdom's gram-negative/gram-positive divide.[6] The clindamycins, fluoroquinolones, and cephalosporins were truly "big guns" that could be deployed in a whatever-might-ail-you fashion. They could be lifesaving in crisis situations, such as when a critical-care doctor lacked the time to run tests to determine exactly what organism was rapidly killing a patient. But that led to the realization, on the part of cost-conscious hospital administrators and harried doctors, that one-size-fits-all antibiotics allowed them to skip the expensive and time-consuming step of identifying infectious organisms, even when they had time to do so. Taking a "what can it hurt?" attitude, surgeons likewise began starting their patients on broad-spectrum drugs before surgery and keeping them on the drugs for days afterward as a hedge against any possible kind of infection. Outside the hospital, general physicians welcomed the convenience and efficacy. "Physicians felt safer with broad-spectrum drugs," Eichenwald recalls, "as in, 'I'm not really sure what bug I'm dealing with here. I should use something that will kill everything.' And the pharmaceutical industry heavily promoted this sentiment in advertising, with words to the effect of 'Be sure!'"

But all this convenience had a dark side. The scattergun attack of a broad-spectrum antibiotic razes not only the disease-causing organism that is its intended target but also the body's trillions of resident bacteria, from potential troublemakers like staph to protective and otherwise beneficial microflora. As a result, these drugs selected for resistance across the spectrum of bacteria that call the human body home. The consequences would not be understood for decades. But by the 1980s, this much was clear: the pace of the Red Queen's treadmill was accelerating beyond the pharmaceutical industry's ability to keep up.

MICROSCOPIC MATING GAMES

That bacteria could evolve and evolve rapidly became painfully obvious almost as soon as the first bacteria-killing medicines came into use: sulfa drugs in the 1930s, followed by penicillin in the 1940s. Within a couple of years of each new drug's introduction, resistant strains of disease-causing microbes appeared. Once they did, their imperviousness to attack seemed to spread rapidly among other kinds of bacteria. Sometimes doctors encountered infections whose cure demanded increasingly higher doses and longer courses of a given antibiotic. Other times they found themselves blindsided by the abrupt appearance of full-blown resistance.

Scientists were flummoxed. Could it be that bacteria gradually developed a tolerance to a drug, the way a person adapts to, say, high altitude or spicy food? The other possibility: within any given colony of bacteria, there might exist one or two microbes with a lucky mutation that allowed them to survive an antibiotic's chemical assault. Once the drug had killed off all the susceptible bacteria, the surviving mutants could multiply and take over—spawning a new strain of drug-resistant clones.

In 1951 the husband-wife team of Joshua and Esther Lederberg, microbiologists at the University of Wisconsin, demonstrated their elegantly simple proof that preexisting mutations, not gradual tolerance, accounted for many instances of new drug resistance. Using swatches of velvet mounted on round wooden blocks, each one the size of a petri dish, the Lederbergs found they could pick up hundreds of pinprick-sized colonies of bacteria from one petri dish filled with bacteria and then deposit them, in the exact same positions, on a series of other dishes infused with an antibiotic such as streptomycin. If resistance stemmed from a preexisting mutation in a few bacteria on the original plate, then colonies would appear in the exact same spots on all the antibiotic-laced dishes. This is exactly what the Lederbergs found, with preexisting resistance to streptomycin turning up once or twice in every 1 to 10 million bacteria cells.[7] This meant that every new antibiotic became a powerful new force for bacterial evolution, winnowing away every bacterium but the otherwise unremarkable one that could survive its effects. With its competition gone, that lucky mutant could repopulate, giving rise to a newly resistant colony overnight.

The news engendered a false sense of security in the medical community. If no more than one in 10 million bacteria could mutate its way into resisting an antibiotic, the solution was simple enough: just deliver a one-two punch, administering two different drugs simultaneously. The chances that a bacterium would be resistant to two different antibiotics were astronomical: one in 10 million times 10 million, or 100 trillion (10^{14}). The problem was, the bacterial world did not appear to be listening. Resistance to multiple drugs had already begun showing up in hospitals and doctors' offices in the late 1940s and early 1950s. In some cases the resistance seemed to build one antibiotic at a time. That is, a strain already resistant to one antibiotic would spin off a substrain that could resist another antibiotic as well. Strangely, some doctors were reporting that they were encountering bacteria that suddenly went from wholly susceptible to standard antibiotics to fully resistant to multiple drugs. Most medical experts dismissed these reports as far-fetched.

Though no one realized it at the time, earlier work by the Lederbergs had already hinted at the many exotic tricks that "simple" bacteria could deploy to evolve around anything that might thwart them. For starters, back in 1946 a baby-faced twenty-one-year-old Joshua had stunned the scientific world by announcing that bacteria have a sex life.[8] The claim seemed outrageous, given that bacteria clearly lack the complicated cell structures that allow larger and more complex organisms to sort and divide their genes into the half-sets contained in eggs and sperm. Everyone knew that bacteria reproduced by simply splitting in two, giving each "daughter" cell an exact copy of the "parent" cell's genes. This simplistic manner of replication allowed a single bacterium to multiply rapidly and exponentially, achieving a population of several million within a few hours, or up to a billion over the course of a day.

Joshua's elegant proof involved a set of mutant *E. coli* bacteria. Each mutant lacked the ability to manufacture one or more different nutrients. So it could replicate only on growth media that were infused with the vitamin or amino acid that it couldn't produce for itself. But by mixing together various combinations of the partially crippled mutants, Joshua produced fully functional "offspring"—that is, bacteria that could make all the nutrients they needed in order to thrive on unsupplemented growth media. Somehow the crippled mutants had combined their genetic assets. In macroscopic life, from worms to humans, parents share their genes through the fusion of egg and sperm. What form "sex" took in bacteria, not even Joshua could say.

Two years later Esther glimpsed part of the "how" behind bacterial sex. She had been working with a fresh set of *E. coli* nutritional mutants when they suddenly lost their ability to "mate." As a geneticist, Esther understood that the disappearance of a trait often provides the key to finding the gene that normally produces it. Over the course of several years she, her husband, and their collaborator in Italy, Luigi Cavalli, would discover a "fertility factor." It sat in a ringlet of genes that floated in the bacterial cell independent of the microbe's main chromosome.[9] Joshua named these circlets of bonus genes "plasmids."[10] It is the job of the plasmid's fertility gene, or F-factor, to direct the cell to build a pilus, a kind of bacterial penis that blebs out from the bacterium's outer membrane to form a bridge with another bacterium and allow the passage of the entire plasmid, along with whatever bonus genetic material it may hold. So blessed, the recipient microbe likewise becomes fertile — that is, able to forge mating bridges and share genes, including those for drug resistance, with still other bacteria.

Around the same time, Esther discovered the first clue to another way that bacteria swap genes. She discovered *lambda*, a bacteria-infecting virus, or "phage," that had been residing silently inside some of the lab's *E. coli* strains for years.[11] On occasion *lambda* popped out of the main chromosome of an infected bacterium to spread to other bacteria. Sometimes the infection killed the new recipients. Other times the phage successfully integrated itself into a bacterium's chromosome and, in this way, was silently passed to future generations.[12] Soon afterward, the third member of the Lederberg lab, twenty-two-year-old Norton Zinder, showed that phages such as *lambda* were in the gene-smuggling business. Specifically, he showed that when a phage cuts itself out of a bacterial chromosome, it occasionally takes a chunk of that chromosome with it. To demonstrate what he thought was happening, Zinder infected colonies of drug-resistant *Salmonella* and then mixed them with uninfected and drug-susceptible strains in a petri dish. In the process, he caught the phages picking up resistance genes and depositing them, in working order, in the next round of bacteria that they infected. Any bacterium that survived the infection found itself instantly drug resistant.[13]

Zinder dubbed this viral gene ferrying "transduction," to distinguish it from "transformation"—yet another way that bacteria can pick up new traits. The discovery of the little-understood process of transformation dated back to a 1928 experiment by the British medical officer Frederick Griffith. Investigating an epidemic of particularly nasty pneumonia, Griffith began comparing and contrasting different strains of the bacterium *Streptococcus pneumoniae*. Among its distinguishing traits, the deadliest strain wore a smooth and slippery capsule that enabled it to escape the grip of the body's immune cells. One day Griffith injected his laboratory mice with a live but harmless strain of *Strep. pneumoniae* along with a bolus of dead cells from the dangerous strain. The mice promptly died. When Griffith examined their blood under his microscope, he discovered that the once-harmless mutants were wearing capsules. Griffith repeated his experiments to ensure that no live, virulent bacteria had slipped through. They hadn't. Somehow the previously harmless, naked streptococcus had picked up the ability to make capsules from the spilled contents of its dead cousin.[14]

In the early 1940s, Rockefeller Institute scientists Oswald Avery, Colin MacLeod, and Maclyn McCarty determined that what Griffith's *Streptococcus* had in fact scavenged from their environment were snippets of the virulent strain's deoxyribonucleic acid, or DNA. In doing so, they provided the first convincing evidence that DNA was the stuff of which genes were made.[15]

When the Lederberg lab published its string of discoveries in 1952, it became clear that bacteria have a plethora of ways to pick up new traits. They can scavenge genes, swap them through conjugation, or have them spliced into their chromosomes by bacteria-hopping phages.[16] We now know of yet a fourth kind of genetic taxi traveling through the bacterial world: so-called jumping genes, or transposons, first discovered by geneticist Barbara McClintock in the 1940s. In essence, a transposon consists of a gene or a group of genes flanked by a pair of "insertion sequences"—segments of DNA that periodically cut themselves out of one chromosome and insert themselves, and all their intervening genes, into another.

In the cells of plants and animals—which contain many chromosomes inside their central nucleus—a transposon simply mixes up the locations of various genes, occasionally turning them on or off in the process. By contrast, in a bacterium, a transposon can move genes from the cell's main chromosome into a plasmid for export to other bacteria. Some transposons even contain genes for independent transport. Once they pop out of the main chromosome, they direct the cell to build a mating bridge over which the transposon slips a copy of itself.

THE BACTERIAL SUPERORGANISM

Nothing suggests that antibiotics created the means by which bacteria can share genes. Samples of bacteria taken from patients in the early 1920s have been found to contain plasmids, phages, and transposons— all poised and ready to move. But these same stored bacterial samples reveal that drug-resistance genes were rare to nonexistent in the bacteria that caused infections in the early twentieth century.

Microbiologists of the 1950s did not appreciate the stunning extent to which bacteria swap genes, but the ramifications were becoming obvious in medical centers around the world. In 1959 Japanese hospitals experienced outbreaks of multidrug-resistant bacterial dysentery. The shigella bacteria, which caused the outbreaks, were shrugging off four different classes of previously effective antibiotics: sulfonamides, streptomycins, chloramphenicols, and tetracyclines. Strangely, the shigella bacteria isolated from patients prior to treatment with antibiotics usually tested fully susceptible to all four classes of drugs. But application of any one of the drugs somehow rendered the bug fully resistant to them all. It made no sense until a team of researchers went back to the stool samples taken from the patients before treatment. Among the patients' normal intestinal flora, the researchers found strains of E. coli that were resistant to all four classes of drugs. Could the E. coli have somehow transferred their multidrug resistance to the far more dangerous shigella? As it turned out, that was exactly what had happened. In fact, the Japanese researchers found it quite easy to transfer multidrug resistance from E. coli to shigella and back again simply by mixing resistant and susceptible strains together in a test tube.[17]

English-speaking microbiologists learned the news in 1961. In the May issue of the widely read Journal of Bacteriology, Tsutomu Watanabe and Toshio Fukasawa, of Tokyo's Keio University, not only relayed their colleagues' findings but reported the fantastical results of their own research. They had documented both plasmids and phages loaded with drug-resistance genes flying between four different species of bacteria—*E. coli, Shigella dysenteriae, Salmonella typhimurium,* and *Salmonella enteritidis.*[18] The latter two rank among the most common causes of food poisoning. Moreover, Watanabe and Fukasawa clocked the amount of time it took to transfer multidrug-resistance genes in the test tube at an average of fifteen minutes. By comparison, bacteria-contaminated food takes hours to pass through *E. coli*'s territory in the human intestines—more than enough time for the bacteria to get intimate and share a few tricks.

Microbiologists reading the report had to absorb the findings in waves: first, they had to accept that bacteria could amass drug-resistance genes like so many charms on a friendship bracelet. Second, any one antibiotic could promote the spread of resistance not just to itself but to a number of other drugs if several resistance genes happened to reside on the same plasmid or were packed into the same phage. For that matter, an antibiotic could promote the spread of any gene—however dangerous or benign—that ended up linked to a resistance gene.[19] Even more profoundly, perhaps, the Japanese had shown that bacteria are not deterred by the species barrier. While salmonella, shigella, and *E. coli* are within the same bacterial family, their exchange of genes crossed not just species but genera. It was the first clear sign that the unfettered exchange of genes, in effect, transformed the entire bacterial kingdom into one huge superorganism, united by the fundamental drive for survival.

DANGER IGNORED

While microbiologists took note, the news of gene-swapping bacteria barely grazed the consciousness of practicing physicians. The specter of insurmountable drug resistance seemed a distant worry, given the seemingly endless supply of newer and ever more powerful antibiotics. Then the antibiotic pipeline began running dry. In the 1980s pharmaceutical scientists realized they were running out of new biochemical targets for antimicrobial drugs. The following decade brought an unprecedented rise in highly resistant nosocomial, or "hospital acquired," infections. Besides the usual troublemaker (staph), these infections also included drug-resistant strains of normal microflora that had rarely caused problems before the advent of antibiotics. Chief among these "good germs gone bad" were the enterococci, a family of orb-shaped bacteria that had ranked among the most innocuous as well as the most widespread of bacteria native to the human intestines. Historically, enterococci did not even cause infections when they contaminated open wounds.

Two inherent traits predisposed the enterococci to cause trouble under the influence of antibiotics. For one thing, they're naturally resilient to life outside the body and can survive in the open air for hours; in addition, they can tolerate low to middling doses of noxious chemicals such as disinfectants and antibiotics. By 1980 two of the hardiest species—*Enterococcus faecalis* and *E. faecium*—had become scourges of the critical-care unit, where intravenous lines, urinary catheters, and ventilation tubes introduced them into the bloodstream and internal tissues of the weakest, most immune-compromised patients.

Initially, doctors assumed that patients were becoming infected by enterococci from their own guts. Soon it became clear that hospitals were breeding their own distinct strains of enterococcal superbugs. A hospital's antibiotic-infused environment favored the emergence of resistance. And imperfect sanitation favored superhardiness—strains able to stubbornly persist for days, weeks, possibly months, on hospital-room surfaces such as bed rails and countertops and on objects such as the stethoscopes, telephones, and beepers that doctors and nurses handled between patients.

Worse, enterococci were proving adept at picking up troublesome genes from other hospital bacteria. Not only did they collect dozens of genes for drug resistance, they acquired DNA for human-cell-destroying toxins called hemolysins and for adhesins, chemical claws that enabled them to climb into previously inaccessible territory such as the urinary tract and the bladder.[20] In the process, the hospital-bred enterococcus became a major repository for dangerous genes that, in turn, spread to other bacteria such as *Pseudomonas aeruginosa* and *E. coli*. Arising from ordinarily harmless stock, these microbes followed enterococcus in becoming major causes of hospital-acquired infections. By the mid-1990s, nosocomial infections were killing an estimated eighty-eight thousand Americans each year, about one every six minutes—more than car accidents and homicides combined.[21]

Most worrisome of all, hospital enterococci had picked up five genes that, together, conveyed total resistance to what had become the antibiotic of last resort—vancomycin. Developed in 1956, vancomycin had remained largely unused, owing to its toxicity and poor absorbability. It tended to cause kidney damage and hearing loss and had to be administered intravenously. But the 1980s brought the rise of hospital strains of staph that were resistant to everything else. Between 1984 and 1994 the use of vancomycin in U.S. hospitals more than quintupled, from 2 to 11 metric tons.[22]

Though the heavy tonnage of vancomycin was aimed at hospital strains of methicillin-resistant *Staph. aureus*, or MRSA, it had the unintended effect of promoting the rise of a dreaded new acronym: VRE. Vancomycin-resistant enterococcus made its U.S. hospital debut in 1988.[23] Within five years it was showing up in nearly one in ten hospital patients, and the majority of those who spent several days or more on vancomycin.[24] Fortunately, VRE caused no harm in and of itself so long as it stayed put in a patient's intestines. The high anxiety centered on the devil spawn that might result when VRE and MRSA cozied up inside the same patient. The reconfigured VRSA could produce wholly unstoppable staph pneumonias, blood infections, meningitis, and more.

Researchers already knew that enterococcus *could* transfer its vancomycin-resistance genes to *Staph. aureus*. They had goaded the exchange in laboratory experiments. They also knew that staph occasionally commingled with enterococcus in the intestines, the staph ending up there when swallowed with some nasal mucus. Still, as the 1990s ticked by with no sign of VRSA, many hoped that some unknown incompatibility was preventing the dreaded mating from ever occurring outside a test tube.

Then, in June 2002, VRSA showed up in a swab taken from the dialysis catheter of a forty-year-old diabetic in Michigan.[25] On several occasions during the previous year, doctors had put the woman on vancomycin, first for her chronic foot ulcers and most recently for a life-threatening MRSA blood infection. The doctors removed the VRSA-contaminated catheter, and the incision where it had been inserted healed without problem. A week later the VRSA showed up again, this time in an infected foot ulcer, from which it could easily enter the woman's bloodstream. Good fortune ruled again: lab tests showed that the otherwise highly drug-resistant strain was susceptible to a thirty-year-old sulfa drug, trimethroprim-sulfamethoxazole (Bactrim). Two more documented cases of VRSA followed in 2004, both in New York State. Disturbingly, one of these initially escaped detection with standard laboratory tests, leaving open the possibility that VRSA might be more widespread than anyone realized.[26]

In the interim, the FDA fast-tracked approval for linezolid, the first in a new class of antibiotics called oxazolidinones. Like many previous antibiotics, oxazolidinones gummed up a bacterial cell's ability to make proteins. However, they did so in a new, potentially more effective way: instead of blocking an intermediate step in the process, linezolid prevented the assembly of the entire protein-making apparatus.[27]

Once again, a drug's manufacturer (Pfizer) claimed it had produced a resistance-proof magic bullet. More cautiously, the U.S. Food and Drug Administration ordered linezolid reserved for treatment of complicated hospital infections that were either suspected or known to be resistant to standard drugs—a move that no doubt reduced its profitability to Pfizer. Regardless, resistance cropped up within a year of linezolid's debut. This time the resistance didn't appear overnight, as with the swap of a plasmid. Rather, it built up over weeks and months in patients on long-term courses of linezolid. In response, doctors administered higher and higher dosages—until the drug stopped working altogether.[28]

Genetic analyses of linezolid-resistant strains of staph and enterococcus have since shown that any one of a dozen or so mutations can render them resistant. Worse, these mutants have proven themselves extremely fit. As of 2005 they were sparsely but widely distributed in medical centers around the world, even showing up in patients who had never taken the drug.[29] In the half century since the introduction of penicillin, the accelerating pace of drug resistance had reversed our early victories and turned the war on germs into a kind of marathon in which our foes were showing strong signs of pulling ahead.

OLD HABITS, NEW INSIGHTS

"A hospital can be a very dangerous place to be sick," says infectious-disease specialist Curtis Donskey. A trim, button-down Midwesterner, Donskey makes the statement without any hint of irony as he sits in his office at Stokes VA Medical Center, in Cleveland. In 1991, as a resident fresh out of medical school, he grappled with one of the nation's first outbreaks of vancomycin-resistant enterococcus, at Brown University's Miriam Hospital. In 1995 he arrived at Cleveland's Case Western University with a grant to study the genes behind VRE's multidrug resistance.

Early in his career, the meticulous young Donskey proved adept at the kind of exacting laboratory work required to dissect resistance genes and track their spread. But his broader, messier interest centered on the mechanisms that allowed this intestinal superbug to jump so readily between hospital patients. Contaminated instruments and hands might spread VRE, but hospital workers rarely became infected; nor did those with whom they came in contact outside medical center walls.

"What we were seeing was far from simple," says Donskey. "Vancomycin-resistant bugs weren't showing up just in patients taking vancomycin. Patients getting other, unrelated antibiotics tended to pick up VRE as well." Donskey began looking at the problem from what he calls "an ecological point of view," focusing on VRE in its natural habitat, the human intestinal tract—in particular, the intestinal tract under the influence of antibiotics.

In his research lab at the VA hospital, Donskey administered antibiotics to mice, cycling each set of animals through one type of drug regimen after another. A few days into each cycle, he gave each mouse a squirt of drinking water laced with VRE. A clear pattern quickly emerged: the VRE consistently colonized mice receiving drugs such as clindamycin or metronidazole, which are known for their effectiveness against anaerobic, or oxygen-shunning, bacteria—the kind of bacteria that predominate in a healthy bowel. By contrast, when Donskey gave the mice antibiotics with little effect on the gut flora (cefepime, aztreonam, and others), the animals quickly eliminated their force-fed VRE.[30]

Donskey's laboratory results correlated exactly with what he was seeing in patient charts at the hospital. "Those on the disruptive antibiotics, in terms of the normal intestinal microflora, were the ones ending up colonized with hospital pathogens, in particular klebsiella, clostridium, and VRE." When he further analyzed hospital records, Donskey saw that once a patient picked up VRE, keeping him or her on an anti-anaerobic drug produced a stunning population spike in the drug-resistant enterococcus. Over the course of a few days, the concentration of VRE in the patient's stool might increase a hundred thousand times over. Such extreme concentrations virtually assured that stray VRE would begin spreading from the patient's stool to objects throughout his or her room—invisibly transferred by touch and movement to other parts of the body, to gowns and linens, to call buttons, bedrails, dishes, table surfaces, and so on. The widespread contamination greatly increased the chances that the bacteria would end up in the patient's bloodstream via a catheter or IV, even as it fueled VRE's spread to other patients via the hands of doctors, nurses, and orderlies.

The good news out of Donskey's study was that when physicians switched a patient to more gut-sparing antibiotics, the normal intestinal microflora quickly rebounded and suppressed the VRE back to trace levels.[31] Blood infections decreased, as did patient-to-patient spread. When Donskey published his results in *The New England Journal of Medicine* at the end of 2000, he effectively secured his promotion to the arguably thankless position of Case Western's chief of infection control—a respected position to be sure, but one that has required him to endlessly hector doctors into curbing their entrenched and dangerous habits.

In a 2001 review of drug-prescribing practices, for example, Donskey found that many M.D.'s were not only choosing unnecessarily disruptive antibiotics, they were using antibiotics when none were needed in the first place. It wasn't that doctors were mistakenly prescribing antibiotics for viral, or nonbacterial, infections. Rather, some were using the drugs in an effort to prevent *possible* bacterial infections, and many more were keeping patients on antibiotics far longer than necessary.[32]

"It was as if, once they started a patient on antibiotics, they were reluctant to take them off," Donskey says. But Donskey knew that using antibiotics as a hedge against infection produced the opposite effect. Crunching the numbers, he showed that every day a patient spent on antibiotics substantially increased the risk that he or she would pick up a VRE or another drug-resistant nosocomial infection. In all, Donskey found that nearly a third of the days that hospital patients spent on antibiotics were unjustified. But if anything, his 2001 report underestimated the problem, Donskey says, for he excluded patients in intensive care units, where doctors must make life-or-death decisions about antibiotics without having the time to find out if the drugs are needed.

Even after decades of being lectured about antibiotic abuse, too many doctors still think of antibiotics as benign and use them with a "what can it hurt?" attitude, Donskey says in frustration. "They know that overuse can promote antibiotic resistance in the long term, but they still don't appreciate the immediate dangers." The bug that finally shook some doctors into awareness made headlines in December 2005, with the report of a national epidemic of a highly drug-resistant and deadly strain of *Clostridium difficile*.[33]

Since the 1970s, doctors and pharmacists have known that *C. difficile* is the most common cause of postantibiotic diarrhea. It began causing problems with the introduction of broad-spectrum antibiotics such as clindamycin. As it turned out, *C. difficile*'s naturally resistant drumstick-shaped spores had always been widespread in nature. Everyone swallows a few now and again. *C. difficile* causes no problems—except when antibiotics disrupt the gut's normally tight-knit community of native microorganisms and so allow the bacterium to take over. Ordinary *C. difficile* weathers a course of antibiotic inside its spores; then, as soon as the person finishes the drug, it germinates and flourishes before the normal residents of the colon have a chance to reestablish themselves. The resident microbes usually win out in the end, eliminating *C. difficile* from the intestines after a few days of intestinal upset. But when they fail to do so, *C. difficile*'s diarrhea-inducing toxins can trigger severe colitis, or inflammation of the intestines.

By the 1980s *C. difficile* spores had become particularly abundant in hospitals, in part because the spores tend to defy eradication by even industrial cleaning products. As a result, some 20 to 40 percent of hospital patients were picking up this bug during their stay, almost always after a course of antibiotics.[34] Still, except on rare occasions, the infections produced nothing worse then temporary diarrhea. Then came the new *C. difficile*.

In the late 1990s hospital patients began dying of severe colitis caused by a strain that had lost the brakes on toxin production. This new strain churned out twenty times more colon-inflaming poisons, in some cases producing deadly perforations that allowed instestinal bacteria to spill into the abdomen and bloodstream. In addition, it had picked up genes for active resistance to the widely used fluoroquinolone antibiotics. This meant it could begin actively overgrowing in the intestinal tract as soon as a person *began* taking one of these drugs, rather than retreating into dormant spores until the drug cleared the patient's system.

The first reported outbreaks appeared in Canada in 2003, when the hypervirulent strain killed more than a hundred patients in Quebec hospitals over the course of six months. Many of the victims had checked in for routine procedures. The next year it killed a staggering two thousand Quebecers, and similar outbreaks were reported in Britain. British tabloids went crazy, running headlines like "Toenail Surgery Nearly Killed Me" and "Hospital Blamed for Mum's Horrible Death."[35]

Meanwhile, in the United States, the Centers for Disease Control and Prevention (CDC) suspected that American hospitals were experiencing similar problems. But because the United States has no nationalized health system or federal laws mandating disease reporting, the CDC had no clear way to gauge what was happening. Case Western's Donskey recalls picking up a Cleveland newspaper in the summer of 2005 and reading headlines still trumpeting "Canada's hospital superbug," with no mention of the growing crisis in the city's own hospitals. "We knew that many hospitals here had the same problem," he says of his discussions with colleagues. "But we had no more than rumors to go on. No real data. No one taking action."

The only way to tease out an approximation of what was going on in the United States was to comb through years of hospital discharge records. The CDC epidemiologist Clifford McDonald headed the herculean effort, reporting the results at the end of 2005. Over the previous five years, colitis-associated death rates at American hospitals had risen from less than 2 percent to as high as 17 percent.[36] Not coincidentally, the rate at which doctors were diagnosing *C. difficile* infections had more than doubled during this time.

At the same time as CDC epidemiologists were sifting through hospital records, Loyola University infectious-disease researcher Dale Gerding was trying to trace the hypervirulent new strain back to its source. Using C. *difficile* samples isolated from patients over the previous twenty-five years, he found the first clear match in an isolate dated 1984. Its source: a hospital patient in the midwestern United States.[37] As the story unfolded, it became clear that not only was the strain born in the United States, but American hospitals had experienced major unreported outbreaks years before those in Canada and the United Kingdom. The first was at the University of Pittsburgh Medical Center in January 2000. Over the following fifteen months, 253 of the center's patients developed severe C. *difficile* infections; eighteen of them died when their infections perforated their colons. Another twenty-six went home with colostomy bags, their large intestines left in the operating room in a race to save their lives.[38] "We had no idea what was going on," the medical center's chief of infection control sheepishly told the media when he reported the outbreak in 2005.[39]

Still the news got worse. In December 2005 the CDC relayed a spate of reports that it had just received of previously healthy children and adults who were coming *into* hospital emergency rooms already infected with the hypervirulent bug. Though the CDC appreciated the prompt reporting on the part of the hospitals, in essence it was as if they were announcing that the barn door was open after a few horses had wandered back home. No longer confined to the hospital, the new C. *difficile* had spilled out into the general population and appeared to be spreading. The CDC's initial announcement described twenty-three of these community cases, with victims ranging in age from six months to seventy-two years and including ten pregnant women, one of whom miscarried twins before dying.[40]

Just as multidrug-resistant staph revealed the hard reality that antibiotics breed resistance, C. *difficile* may finally be awakening doctors to the danger inherent in razing the body's "good bugs" along with the bad, says Donskey. "No one is saying, 'Don't use antibiotics,'" he emphasizes. "But like most drugs they have to be used carefully, in ways that minimize harm."

OUT OF THE HOSPITAL AND INTO OUR DAILY LIVES

The 2003 flu season started early in North America, with the first cases showing up in the fall. By Thanksgiving doctors were seeing the usual flu-related pneumonias. As always, the most severe cases resulted from secondary bacterial infections in flu-congested lungs. Most of these infections responded quickly enough to a standard antibiotic such as amoxicillin, or perhaps a bigger gun such as azithromycin or cephalexin. Doctors knew that they would lose a few of their most frail, elderly, or immune-compromised patients. But then the young and strong began dying.

John Solomon Francis, a thirty-three-year-old infectious-disease researcher, saw some of the first "atypical victims." A burly six-foot-four native of Queens, New York, Francis normally defused his intimidating outward appearance with a broad, gap-toothed smile. But on a late-November day in 2003, he stood grimacing at the bedside of a thirty-one-year-old woman lying unconscious in the intensive care unit of Johns Hopkins Bayview Medical Center, in Baltimore. As the woman's chest rose and fell with the wheezy regularity of a mechanical ventilator, Francis flipped open her chart and reviewed her intake information. No prior hospitalizations. No underlying medical conditions. A friend had half carried the woman into the emergency room a few hours before. The intake nurse noted shortness of breath, coughing, and hemoptysis, or blood-streaked sputum. Before she passed out, the woman had said she'd caught the flu from a friend the previous week and that she'd felt so awful that her family doctor had given her an antibiotic, Zithromax.

That should have been more than enough to stop any bacterial infection, Francis thought. But the woman was clearly suffering from severe bacterial pneumonia. Now it was Francis's job, as the infectious-disease consultant on call, to recommend treatment to the attending physicians at the woman's bedside. He grimly recommended an intravenous cocktail of vancomycin, gatifloxacin, and meropenem. It was overkill, perhaps, but there was no time to wait for laboratory cultures to determine what was killing this young woman.

Replacing the chart at the end of the woman's bed, Francis headed downstairs, where a radiologist was pulling up images from a CT scan of the woman's chest. Francis could see the gaping holes, some as large as an inch square, that riddled what should have been smooth lung tissue. The radiologist pointed out the haziness that suggested the holes were abscesses filled with blood and pus. A thick white halo surrounding the lungs showed the pleura, or lining of the chest cavity, to be likewise filled with fluid.

"If she was an injection drug user, I'd think endocarditis," Francis offered, referring to the scenario of a dirty needle introducing bacteria into the bloodstream, infecting the heart, and from there spreading to the lungs. Walking back toward the ICU, Francis found the woman's parents and asked the hard questions. No, they said, she had done nothing even remotely related to drug abuse. She was a light smoker, yes, maybe six cigarettes a day. Deepening the mystery, the woman's blood and sputum cultures came back the next morning positive for methicillin-resistant *Staph. aureus*—a bane of hospital life to be sure, but something seldom brought *into* a hospital. Stranger still, though in a good-news sort of way, the bug tested susceptible to several antibiotics that seldom worked against hospital MRSA.

Francis saw the second case of overwhelming pneumonia a few days later, this time in a previously healthy fifty-two-year-old man who was likewise suspected of having had the flu. There was no need to rush over, the pulmonologist told Francis when he took the late-night call. They would start the patient on the same trio of intravenous antibiotics they'd given the young woman. But when Francis stopped by first thing in the morning, the patient was dead. The just-arrived lab report showed blood cultures positive for MRSA with the same pattern of limited drug resistance as the woman still on life support down the hall.

That evening Francis described the two cases to his supervising physician, Eric Nueremberger. The following morning Nueremberger handed Francis a journal article he'd recently read about a spate of fatal staph pneumonias among children and young adults in France.[41] The French researchers linked the infections to an aggressive new strain of staph that carried a gene for a necrotizing, or tissue-killing, toxin: Panton-Valentine leukocidin, named after its discoverers and its ability to punch holes into the infection-fighting immune cells known as leukocytes, or white blood cells. As at Johns Hopkins, the French doctors had confirmed that flu, or flu-like, illness had preceded the deadly infections in the vast majority of their patients. In the conclusion of their paper, the French researchers compared the swift and terrible deaths with those seen during the Spanish flu pandemic of 1918. Francis recalled one of the classic accounts he'd read in medical school:

> As their lungs filled . . . the patients became short of breath and increasingly cyanotic. After gasping for several hours they became delirious and incontinent, and many died struggling to clear their airways of a blood-tinged froth that sometimes gushed from their nose and mouth. It was a dreadful business.[42]

A dreadful business indeed. The flu virus circulating that fall—a Fujian strain little different from the "Panama flu" seen the previous year—was no worse than most. Yet Francis's first patient was barely off life support when two more flu victims, both women—one twenty, the other thirty-three—showed up at Johns Hopkins Hospital. Both lay unconscious on life support by the time their blood cultures came back positive for MRSA, and both suffered the kind of lung-shredding pneumonia that Francis had just witnessed for the first and second time. Worse, within days their infections progressed to overwhelming sepsis, or blood infection; their immune systems' frenzied but ineffective response was to produce massive clotting. As a result, both women suffered multiple strokes and loss of circulation to their extremities. Their toes and fingers blackened. Infusions of anticlotting drugs saved their fingers, but doctors had to amputate the twenty-year-old's left leg below the knee. The thirty-three-year-old lost both legs. Meanwhile, Francis's first patient was able to return home for Christmas, this after four weeks in intensive care.

During this time Nueremberger and Francis began spending nights and weekends in the hospital's research lab, testing for the presence of the Panton-Valentine leukocidin gene in the staph cultured from the four patients. They found it in all of them. Further gene profiling showed the isolates to be the exact same strain. When Francis contacted the CDC, he learned that Johns Hopkins was not alone. Reports of similarly virulent and drug-resistant staph pneumonias were trickling in from across the country, the majority involving previously healthy children and young adults and always coming on the heels of a viral respiratory infection such as the flu.

Though these MRSA victims ended up in the hospital, where several died, all had picked up their staph in the outside community. Just a few years earlier, CDC epidemiologists had skeptically set aside reports of so-called community-acquired MRSA pneumonia. "We made the mistake then of assuming that those early reports involved MRSA that had some kind of hidden connection to a health care setting," admits the CDC's Jeffrey Hageman.

Those first accounts came out of Chicago's Wyler Children's Hospital, where in 1996 infectious-disease consultant Betsy Herold began seeing otherwise healthy children arriving in the emergency room with MRSA infections in their bones, soft tissue, and blood. Herold asked research fellow Lilly Immergluck to review hospital laboratory records of antibiotic resistance for all staph infections treated over the previous three years. Immergluck found fifty-two cases of MRSA, thirty-five of them in previously healthy children. The bacteria infecting the thirty-five kids all shared the same atypical resistance and susceptibility pattern that Herold saw in 1996. For comparison, Immergluck reviewed hospital records for 1988 through 1990. She found just eight cases of this distinct strain, which would later become known as community-acquired MRSA.

Most worrisome of all, between 1990 and 1995 the rate of community-acquired MRSA at Chicago Children's had increased from 10 per 100,000 admissions to over 250. When the *Journal of the American Medical Association* published Herold's findings in 1998, those who took it seriously—primarily doctors on the South Side of Chicago—presumed that a feral strain of hospital staph was present in their urban population, one that had dropped a few resistance genes in the less-antibiotic-infused neighborhoods outside their medical-center walls.[43]

Then in early 1999 state epidemiologists in North Dakota and Minnesota reported the deaths of four children from overwhelming MRSA infections: a seven-year-old African-American girl who had arrived at the downtown Minneapolis Children's Hospital with excruciating pain tracing to a hip infection; a sixteen-month-old baby spiking a fever of

105°F when her parents rushed her to the Indian Health Service clinic near their North Dakota reservation; a thirteen-year-old girl in rural Minnesota whose MRSA rapidly spread from her lungs to the lining of her brain; and a North Dakota toddler who died within twenty-four hours, even though emergency room doctors immediately suspected MRSA and started treatment with vancomycin. These four deaths all involved drug-resistant staph in previously healthy children with no known contacts to the usual MRSA haunts of hospitals and nursing homes.[44] Though still widely regarded as flukes, the fatalities prompted more researchers to look at the unusual cases in their own communities.

In Houston, pediatrician Sheldon Kaplan, of Texas Children's Hospital, had been seeing MRSA-infected children arrive in his hospital emergency room with increasing frequency—lung, bone, joint, soft tissue, and blood infections. "When I'd report these cases, everyone said there had to be a connection to someone with a hospital infection, if only a visit to a nursing home or a relative who worked in the hospital." But the more Kaplan investigated, the more he became convinced that these young patients were not luckless victims of an escaped hospital bug.

"For one thing, their MRSAs didn't have the same resistance profiles as our hospital strains," he says. The methicillin-resistant staph coming *into* Texas Children's Hospital proved resistant to the penicillin and methicillin family of drugs, but little else. What these community strains lacked in resistance, however, they more than made up for in virulence. Something about these new bugs allowed them to cause far more serious, invasive disease than any strain of staph Kaplan and his colleagues had ever seen.

Like Herold before him, Kaplan assigned his postdoctoral fellows the task of strain-typing the laboratory culture of every staph infection that came into the hospital from the outside community. In February, the first month of their study, Carlos Sattler and Edward Mason discovered that MRSA already accounted for more than *one-third* of the community-acquired staph infections being seen at Texas Children's. Within a few months, MRSA was actually surpassing "normal," or methicillin-susceptible, staph as the leading cause of serious staph infections among Houston's children.[45]

Meanwhile, across the nation, outbreaks of "flesh-eating" skin infections began spreading among athletic teams and other groups in close physical contact—jail inmates, residents of homeless shelters, gay men, and preschoolers. Most famously, the 2003 football season brought outbreaks among a half dozen turf-burned linemen and linebackers with the St. Louis Rams, who spread it to opposing players with the San Francisco 49ers.[46] In Miami the same year, Dolphins linebacker Junior Seau and kick returner Charlie Rogers ended up in the hospital on intravenous vancomycin to stop the limb-threatening damage, as did Tampa Bay Buccaneer Kenyatta Walker and the Cleveland Browns' Ben Taylor, who required emergency surgery.

Genetic analysis undertaken by the CDC revealed that the strain of staph causing all these infections shared a relatively small "cassette" of resistance genes that rendered them impervious to beta-lactam antibiotics, such as the penicillins, methicillins, and cephalosporins, but also left them vulnerable to many second-line drugs to which hospital strains of MRSA had long ago grown resistant. This "resistance lite" came packaged with an array of genes for increased invasiveness of internal tissues, evasiveness of the body's immune soldier cells, and sheer toxicity. The best studied of these so-called virulence factors has been Panton-Valentine leukocidin (PVL). The possibility remains that the PVL gene is simply a telltale marker for a host of other, as-yet-undiscovered virulence genes. In any case, its presence clearly correlates with staph infections that produce massive tissue destruction and an overwhelming yet ineffective immune response that compounds the crisis with septic shock—a collapse in blood pressure that can result in rapid organ failure and death.

Some older physicians were not surprised when a recent analysis of fifty-year-old cultures of 80/81 staph revealed that it, too, carried the PVL gene.[47] Though 80/81 appeared to fade away with the introduction of methicillin in the 1960s, true to Eichenwald's predictions, it had continued to circulate beneath the medical community's radar, possibly sharing its virulence genes with other strains. Today the predominant subtype of community-acquired MRSA in Europe and Australia turns out to be a direct descendant of 80/81. In the United States, the PVL gene has ended up in an even *more* aggressive staph strain, USA300, and its slightly less deadly cousin USA400. It was USA400 that showed up in Chicago and the Midwest in 1996. And it was USA300 that killed Francis's patients, as well as college student Ricky Lannetti and at least two dozen others in the flu season of 2003–2004. Since then it has become the predominant cause of staph infections seen in hospital emergency rooms across the country—both nasty skin infections and the often-deadly staph pneumonias that have accompanied each year's flu season since 2003.[48] "I hate to think what will happen when we have a truly bad flu virus on our hands," says John Hopkins's Francis. "Whenever I hear people worrying about pandemic bird flu, it's the setup for USA300 that scares me."

THE RESERVOIR WITHIN

For most of us, MRSA and *C. difficile* remain scary but distant threats, random killers to be dodged, perhaps with the magic talismans of antibacterial soaps and gels. More worrisome is that, over the last half century, the typical human body has become a vast reservoir of drug-resistant bacteria. On occasion these resident bacteria cause direct harm, as when they stray from their assigned places to cause, say, a toddler's drug-resistant ear infection or a woman's resistant urinary tract infection. Of greater concern is the ready store of drug-resistance genes that these resident bacteria make available to their far more dangerous kin.

Early inklings of this growing internal danger came in the 1960s, in the wake of Tsutomu Watanabe's discovery that intestinal bacteria can swap some sort of mysterious "resistance factors," which turned out to be plasmids loaded with genes for antibiotic resistance.[49] In 1969 Ellen Moorhouse of Ireland's Royal College of Surgeons reported her discovery of drug-resistant bacteria in the stool of more than 80 percent of one hundred healthy Dublin babies and toddlers. Many of these microbes proved impervious to three or more antibiotics that should have stopped their growth if not killed them outright.[50] In 1972 British bacteriologist Karen Linton followed up on Moorhouse's surprising discovery to find that most healthy people in and around Bristol likewise carried drug-resistant intestinal bacteria, with the rate on average higher among children (67 percent) than adults (46 percent) but highest of all among people who worked with livestock (79 percent). Disturbingly, Linton found that the resistant bacteria she isolated from her volunteers could readily share their resistance genes with drug-susceptible bacteria.[51]

More reassuring was an Israeli study around the same time: microbiologist David Sompolinsky found that while a course of antibiotics produced a dramatic spike in drug resistance among a patient's microflora, resistance levels gradually fell off over the following months. Specifically, after a two-week course of tetracycline or chloramphenicol, more than 80 percent of his volunteers' intestinal bacteria had become resistant to the drugs. Two months later that rate was down to 10 percent—except among hospital patients, who for little-understood reasons maintained their high levels of resistant microflora even after discontinuing antibiotics.[52]

Sompolinsky's findings reinforced the widely accepted view that drug-resistant bacteria were inevitably weakened by their extra baggage of resistance plasmids. Withdraw the selective pressure of antibiotics, the reasoning went, and the resistant bugs would quickly yield to their drug-susceptible kin. That reasoning was enough to assure physicians of the 1960s and 1970s that antibiotic susceptibility could always be restored by giving a patient a "drug holiday" or simply switching to a different antibiotic.

One young physician who was not at all reassured was Stuart Levy. Newly graduated from the University of Pennsylvania Medical School, Levy ranked among a handful of practicing doctors who understood the ramifications of the gene shuffling that was being discovered by bacterial geneticists such as Watanabe, with whom Levy had studied during a medical school semester in Tokyo. In 1977, by which time Levy was working at Tufts University, he began the first large-scale assessment of drug-resistance levels in intestinal bacteria, using a diverse cadre of several hundred volunteers that included hospital patients and university students as well as families from both cities and rural areas in several states. The results shocked even Levy. Drug-resistant microbes made up 50 percent or more of the intestinal bacteria in more than one-third of his healthy volunteers—none of whom had taken antibiotics in more than six months. By Levy's calculations, the average American was excreting 10 million to a billion drug-resistant

E. coli each day.[53] "I can remember my frustration at medical micro-biology meetings," Levy recalls three decades later. "People in the audience would say, 'These resistance factors may be interesting to geneticists, but they're not a clinical problem.' That was very striking to me, because I could see this was just the tip of the iceberg."[54]

To those that heeded them, Levy's studies dispelled the idea that resistance genes were eliminated once a patient stopped taking antibiotics. Other researchers confirmed this worrisome new reality: in the early 1990s Abigail Salyers, of the University of Illinois at Urbana-Champaign, began tracking the rise of a whole suite of drug-resistance genes that she found lurking in many strains and species of *Bacteroides*, one of the intestine's most predominant groups of bacteria. Where did the genes come from, and what drove their spread? she wondered. To find the answers, she became a microarcheologist, using DNA probes to search for the resistance genes in frozen stool samples taken both from healthy volunteers and from hospital patients between the late 1960s and early 1990s.

She found that before the 1970s, the tetracycline-resistance gene *tetQ* showed up in less than a quarter of the *Bacteroides* in her samples. By contrast, more than 85 percent of the *Bacteroides* isolated from her 1990s-era samples carried *tetQ*, regardless of whether the sample came from a hospital patient or from someone who had not used antibiotics in years. Over the same three decades many strains of *Bacteroides* likewise picked up three genes—*ermB*, *ermF*, and *ermG*—conveying resistance to the newer antibiotic erythromycin; their prevalence increased from less than 2 percent in 1970 to over 20 percent in the mid-1990s.[55] Salyers's results provided clear evidence that the prevalence of antibiotic-resistant bacteria in our bodies was building inexorably decade by decade.

With *Bacteroides* making up as much as a quarter of a person's intestinal bacteria, Salyers appreciated the danger of their steadily increasing resistance to two of the world's most commonly prescribed antibiotics. "These bacteria would pose an immediate danger should trauma introduce the microbes into the abdominal cavity," she says, "but the greater danger is that most of us may be walking around with way more than enough resistance genes to turn a previously treatable infection into an unstoppable one." In theory at least, *Bacteroides* could share its genes with dangerous intestinal pathogens such as the food contaminants salmonella, shigella, and campylobacter and with respiratory pathogens such as *Staphylococcus aureus* and *Streptococcus pyogenes*, which regularly pass through the intestines in swallowed mucus and saliva.

Salyers calculated that such bacteria spend twenty-four to forty-eight hours commingling with the "locals" as they pass through the gastrointestinal tract. "And that's more than enough time to pick up something interesting in the swinging singles bar of the human colon," she quips. "Or to view it another way, we've turned our intestines into the bacterial equivalent of eBay. Instead of having to develop drug resistance the hard way—through mutation—you can just stop by and obtain a gene that has been created by some other bacterium." Salyers has already clocked strains of *Bacteroides* excising and transferring their *tetQ* genes among themselves in as little as a few hours. More recently, her research team at the University of Illinois has been testing how far and wide *Bacteroides* can spread these dangerous genes and what conditions goad it to do so.

Indeed, to work in Salyers's laboratory is to play matchmaker to some unlikely couples. Standing at her lab bench, doctoral student Kaja Malanowska lifts the cover from a petri dish to pick up half a billion or so *E. coli* with the tip of a sterile inoculating needle. She adds them to a test tube containing about as many cells of *Bacteroides thetaiotaomicron* floating in a tiny puddle of antibiotic-laced broth.[56] After purging the tubes of oxygen, Malanowska leaves the bacteria to commingle for the night inside the cozy warmth of the laboratory's walk-in "gut," a room-size incubator set to a healthy intestinal temperature of 100°F (37°C). What Malanowska will find a couple of days later is no secret to her mentor. "In the bacterial world's version of casual sex, *E. coli* and *B. theta* will swap a few genes," Salyers explains. Such a mating involves two very distantly related species, akin to a cow mating with a cougar or even a snake.

Even more of a stretch, Salyers found genetic evidence that the gram-negative *Bacteroides* family originally picked up some of these resistance genes at the opposite end of the bacterial kingdom—among gram-positive organisms such as the spore-forming intestinal bug *Clostridium perfringens* and the rod-shaped soil bacterium *Bacillus sphaericus.* Such a gene transfer—from a rigid and heavily armored gram-positive cell to *B. theta's* slippery gram-negative capsule—had long been thought improbable if not impossible. "Rather like an armadillo mating with a squid," Salyers says. Or, in terms of evolutionary distance, more like an armadillo coupling with a sequoia.

Salyers knows that the antibiotic in her student's test tube will not simply fail to kill the bacteria floating inside it but will actually stimulate their DNA transfers. "If you think of transfer of genes between species as the bacterial version of casual sex between strangers, you have to think of antibiotics as their aphrodisiacs," she says. "Tetracycline really turns these guys on." That response reflects what may be a range of cooperative survival techniques employed across the bacterial kingdom. "I'll share my genes if you'll share yours" might be the theme, says Salyers.

Salyers shares the concern of Tufts University's Levy that not just antibiotics but many other modern-day antibacterial products may promote the spread of resistance through these and other bacterial tricks. "Stuart Levy brings up a serious reason for concern," she says, "as some of these products may select for mutations that make bacteria resistant not only to them but to certain other antibiotics." Of particular concern to Levy has been triclosan, the chemical most commonly added to antibacterial soaps, toothpastes, mouthwashes, and household cleaning products. In 1998 Levy lab member Laura McMurry showed that triclosan worked more like an antibiotic than a kill-all disinfectant such as bleach or alcohol. Specifically, she showed that it blocked an enzyme that bacteria use to synthesize fats, an action that *E. coli* and many other microbes could circumvent with a relatively small tweak, or mutation, of their DNA.

Since then Levy's protégés have likewise shown that triclosan can trigger multidrug resistance in *E. coli*, *salmonella*, *shigella*, and other intestinal bacteria. It does so by tripping a genetic master switch, dubbed the multiple-antibiotic-resistance, or *mar*, operon. The *mar* switch, in turn, activates an array of some sixty different survival genes, including one for a so-called efflux pump, which expels not just triclosan but an array of antibiotics from the bacterial cell. The same kind of generic "vomit response" gets triggered when these bacteria encounter disinfectants such as pine oil or chemical preservatives such as benzalkonium chloride and other quaternium ammonium compounds commonly used in eyedrops, nasal sprays, and cosmetics. Consequently, these household products may promote multidrug resistance by selecting for bacterial mutants that keep their antibiotic bilge pumps running round the clock.[57] Potentially compounding the problem, triclosan and its close chemical cousin triclocarban stubbornly persist in treated wastewater and over the last twenty years have become pervasive in groundwater, well water, and freshwater lakes and streams across the United States.[58]

Levy and Salyers are far from alone in churning out research showing that, in a little over half a century, the heavy use of antibiotics and other antibacterial chemicals has transformed our inner microbial ecosystems. Researchers at London's Eastman Dental Institute, for example, recently found that virtually all early grade-schoolers harbor a mouthful of tetracycline-resistant bacteria—even though doctors don't even give tetracycline to children under twelve because the antibiotic stains emerging teeth.[59]

What these studies don't begin to address is the origins of these problematic genes. In rare cases, a new type of drug resistance arises through random mutation. A lucky mutation can change a drug's biochemical target so that the antibiotic can no longer attach itself inside a bacterial cell. A simple mutation can likewise flip the switch that keeps an efflux pump running overtime. But an efflux pump itself constitutes a fully functional biochemical machine whose genetic blueprint took eons to evolve. The same is true for the elaborate genes behind bacterial enzymes such as beta-lactamase, which chop up, block, or otherwise neutralize scores of important antibiotics. Clearly, these resistance mechanisms didn't evolve in the last sixty-plus years. Just as clearly, before the advent of antibiotics, they were rare to nonexistent in the kinds of bacteria that colonize or infect the human body. As it turns out, they were never farther away than the dirt beneath our feet.

RESISTANCE BY THE SHOVEL

As the head of McMaster University's Antimicrobial Research Center, in Hamilton, Ontario, Gerry Wright has the most tricked-out laboratory a drug designer could want, including a $15 million high-speed screening facility for simultaneously testing scores of potential drugs against hundreds of bacterial targets. Yet Wright has found that twenty-first-century technology pales in comparison to the elegant antibiotic-making ability that he witnesses in a clod of dirt.

"It's taken the best minds in synthetic chemistry years of incredible effort to make even small amounts of structurally complex antibiotics such as vancomycin," he explains. "But many kinds of bacteria can do it easily." Of particular interest to Wright and his McMaster team is *Streptomyces*, an expansive genus of soil bacteria that have long fascinated scientists with their intricate colonies of long filamentous cells and their almost fruitlike spore-bearing stalks. On a practical level, these antibiotic-producing bacteria have stocked our medical arsenal with more than a dozen classes of drugs, including the streptomycins, tetracyclines, neomycins, clindamycins, erythromycins, and vancomycins.

In the underground microbial kingdom, these biochemicals appear to play two distinct roles. Studies suggest that, at low concentrations, they function as signaling molecules, allowing bacterial cells to sense and respond to other organisms of their own and other species.[60] At higher concentrations they may play the more familiar role of antibiotic-as-poison, edging back the competition in the endless jostling that takes place in the complex microbial communities that imbue sand and soil everywhere, from the world's deserts to its mountaintops.

Wright began studying *Streptomyces* genes in the mid-1990s for the express purpose of learning some new tricks for drug development. In particular, he and doctoral student Christopher Marshall focused on a stretch of chromosome belonging to the bacterium *Streptomyces toyocaensis* that is known to be involved in the production of teichoplanin, an antibiotic closely related to vancomycin. Their analysis produced a catalog of several dozen genes that included an unexpected bonus—a suite of self-preservation genes that protected *Streptomyces toyocaensis* from its own poison.

It made sense to Wright that the microbe intermixed these resistance genes with the genetic blueprints for its poison-producing machinery. In this way it could efficiently coordinate the manufacture of its "antidote" on an as-needed basis. What Wright did not expect was that the cluster of five resistance genes that he and Marshall pulled out of the dirt microbe seemed eerily familiar. Any microbiologist working with hospital superbugs such as vancomycin-resistant enterococcus would have instantly recognized them: one gene to snip off the antibiotic's binding site in a gram-positive cell wall, two more to produce a drug-resistant replacement part for the deleted chink in the wall, and a final couplet of regulatory genes to turn on the first three, as needed, whenever vancomycin or any of its chemical cousins turned up in the neighborhood.

Wright and Marshall used DNA probes to troll for these same resistance genes among other streptomyces bacteria. They found the genes in the vancomycin-producer *Streptomyces orientalis*, as well as in a half dozen other strains and species that produced chemically related antibiotics.[61] "That gave us a forehead-slap moment," says Wright. "If we had only done this experiment fifteen years ago, when vancomycin came into widespread use, we would have understood exactly what kind of resistance mechanisms would follow the drug into our clinics and hospitals."

With streptomyces bacteria ubiquitous in soil, Wright wondered what else he might find in a shovelful of dirt. His next step, he says, was embarrassingly simple for a scientist surrounded by multimillion-dollar technology—it was "all done with stuff you could have used a hundred years ago," he says. Whenever Wright attended a scientific conference or just went hiking in the woods with his children, he'd bring back a Ziploc bag filled with soil—be it the leafy loam of a forest or the cigarette-strewn flowerbed outside a convention center. He'd then have his students screen the samples for streptomyces and test the bacteria's susceptibility against a gauntlet of twenty-one different kinds of antibiotics. Wright likewise stuffed his students' backpacks with plastic bags, sending them home for the holidays with requests to bring them back filled with soil. Over the next two years, the lab amassed a collection that included Saskatchewan prairie loam, Toronto back-lot clay, fertilized orchard soil from Niagara, and a chunk of the Canadian Rockies. Wright's younger brother, a provincial police officer stationed on the Ontario-Manitoba border, even mailed in a thawed slice of the northern frontier. "We had dirt from Vancouver to Halifax," says Wright.

Students Vanessa D'Costa and Katherine McGran did the laboratory grunt work of isolating the filamentous, spore-forming *Streptomyces* in the accumulated soil. They built up a library of nearly five hundred strains and species, including many never before isolated. Most impressively, all the organisms proved to be resistant, not only to their own signature antibiotics, but to many others. Without exception, each could digest, deactivate, block, expel, or otherwise neutralize multiple antibiotics. On average, they could resist seven or eight drugs apiece, and many could shrug off fourteen or fifteen. In total, the researchers encountered resistance to every one of the twenty-one antibiotics they tested, a gauntlet that included old standards such as tetracycline and erythromycin, as well as such potential new power drugs as the VRE-buster Synercid and the MRSA hopefuls Tygacil and Cubicin.[62] Even more surprising, many of the microbes proved resistant to wholly synthetic drugs such as Cipro and the highly touted new wonder drugs Ketek and Zyvox[63]—chemicals unlike anything the organisms would have ever encountered in nature.

In 2006 Wright published his team's findings in the high-profile journal *Science*, under the headline "Sampling the Antibiotic Resistome."[64] While many scientists expressed surprise that bacteria could nullify more than a dozen synthetic and semisynthetic drugs, Wright says he expected as much. "You can become resistant to antibiotics in many ways, and a lot of them tend to be nonspecific. Efflux pumps, for example, will bail most anything greasy out of a cell." Of special interest, Wright's team encountered several previously unknown resistance mechanisms. Over half the bacteria made a novel enzyme that destroyed both Synercid and Rifadin, the latter a linchpin in tuberculosis treatment. Others neutralized Ketek using the never-before-seen trick of attaching a sugar molecule onto the drug's chemical skeleton to block its antibacterial action.

The report spurred tabloidesque headlines ("Superbugs Abound in Soil!") and much hand-wringing.[65] Forget the prudent use of antibiotics— have we all been tracking the agents of drug resistance into our homes and hospitals on the treads of our shoes? Far from it, Wright concluded. "These genes clearly didn't jump directly from streptomyces into disease-causing bacteria," he explains. For starters, antibiotic-producing organisms such as the *Streptomyces* keep these "suicide prevention" genes securely embedded in their main chromosome. The genes must get repackaged into mobile elements such as plasmids and transposons before they can become part of the bacterial kingdom's gene-swapping network. Second, though the resistance genes that Wright found in dirt microbes closely matched many of those found in infectious bacteria, subtle spelling variations in the genes' DNA told him that they had been passed through intermediaries on their way from one group to the other.

As in a game of telephone, each time a gene gets passed between microbes, subtle variations develop that reflect the DNA "dialect" of its new host. Specifically, the DNA of any organism—from a bacterium to an elephant—has a characteristic "GC content," the proportion of its DNA letters that are made up of the nucleotides guanine and cytosine (which together with adenine and thymine make up the entire DNA alphabet). As a group, *Streptomyces* have some of the highest GC content in the bacterial world. Guanine and cytosine make up over 70 percent of the nucleotides in *Streptomyces* DNA, including that of their resistance genes. By contrast, when these same genes show up in infectious bacteria such as *Enterococcus* or *Staphylococcus*, they have a GC content closer to 50 percent. That's still higher than the genes "native" to these disease-causing organisms. Staph and enterococci both have GC contents around 37 percent. So it suggests that these genes have spent time in intermediary organisms, with their GC content gradually drifting in the direction of each new host.

"What we've got here is like an incomplete fossil record," Wright says. "Now we have the beginnings: the naturally resistant organisms in the dirt. And we have the end: vancomycin-resistant enterococcus in the hospital. They possess the same resistance genes. But we know there's this whole series of missing links involved in how they got from the beginning to the end."

Where to begin looking for those intermediary steps? It is clear that the heavy use of antibiotics in medicine plays a large role. In addition, the heavy use of antibiotics in agriculture—most especially in livestock feed—may have opened a conduit that leads directly to the dinner plate.

DOWN ON THE FARM

> ### "WONDER DRUG" AUREOMYCIN FOUND TO SPUR GROWTH 50%
>
> SPECIAL TO THE NEW YORK TIMES
> PHILADELPHIA—The golden-colored chemical aureomycin, life-saving drug of the group known as antibiotics, has been found to be one of the greatest growth-promoting substances so far discovered, producing effects beyond those obtainable with any known vitamin.[66]

On April 9, 1950, Lederle Lab chemists Thomas Jukes and Robert Stokstad announced their accidental discovery of a "spectacular" new role that antibiotics could play. It was a phenomenon, they believed, that would "hold enormous long-range significance for the survival of the human race in a world of dwindling resources and expanding populations." Just five pounds of unpurified antibiotic, added to a ton of animal feed, increased the growth rate of piglets by 50 percent; similar results occurred in chicks and calves, making them fatter faster than anything in the history of animal husbandry.

The growth-promoting properties of antibiotics had revealed themselves to the Lederle chemists while they were extracting B-12, a vitamin already being used to spur livestock growth, from the wastewater of the company's antibiotic fermentation process. Specifically, they were extracting the B vitamin from vats of *Streptomyces rimosus*, the golden-hued soil bacterium that excretes aureomycin. To their surprise, Jukes and Stokstad found that the unprocessed wastewater sped animal growth to a far greater degree than did B-12 alone. Feeding the animal pure aureomycin produced even more spectacular results. Another Lederle scientist, Benjamin Duggar, had isolated aureomycin just two years earlier, developing it into the broadest-spectrum antibiotic ever—able to eradicate more than fifty kinds of disease-causing organisms known to plague man. But given the potential demand from U.S. farmers, Jukes and Stokstad's discovery had the potential to produce even greater profits for Lederle.

Raising young farm animals on antibiotics didn't result in larger adult livestock. Rather, by speeding their growth, it dramatically cut the time—and therefore the cost—of bringing them to slaughter. At the same time, the modern trend toward large-scale "warehouse farming" was making it impractical to treat sick livestock individually. With the cost of antibiotics falling to as little as thirty to forty cents a pound, it made far more sense to simply treat entire herds and flocks by dumping the drugs into feed and water.

During the press conference announcing their discovery, Jukes and Stokstad assuaged concerns that antibiotics fed to animals would end up on the dinner table. They explained that the chemicals would break down during the course of animal digestion. One indication otherwise earned brief mention the following month. Buried halfway down the "Notes on Science" column of *The New York Times*, it read:

> Wisconsin cheesemakers have had trouble with milk in their vats during the curding process. The trouble has been traced to cattle that were being treated with penicillin or aureomycin for mastitis, an udder disease. Dr. W. V. Price of the University of Wisconsin found that the drugs halt the growth and upset the normal bacteria necessary for good cheese production.[67]

Today U.S. livestock consume over 20 million pounds of antibiotics each year, according to the Animal Health Institute, the lobbying group representing the veterinary drug industry in the United States.[68]

Estimates by the Union of Concerned Scientists, a consumer advocacy group, put the total higher, at nearly 25 million pounds, compared to 3 million pounds used in human medicine in this country each year.[69] Of special concern to the union's scientists is the tonnage fed to animals at low, subtherapeutic doses solely for growth promotion.[70] By their estimate, such nontherapeutic use accounts for 70 percent of all antibiotics fed to livestock, or more than 17 million pounds a year. The industry breaks things down somewhat differently, tallying 2 to 3 million pounds of antibiotics a year for growth promotion and another 20 million pounds or so to prevent infections at times of known stress (when piglets and calves get weaned and when flocks and herds get transported, combined, and crowded) and to stop disease spread when a farmer notices a snickering hen, a runny-nosed calf, or some other sign of individual illness. A little less than half of this total—around 10 million pounds—entails ionophores and arsenicals, antimicrobials not used in human medicine. The remaining 12 to 14 million pounds come out of the same classes of antibiotics used by people—tetracyclines, cephalosporins, fluoroquinolones, penicillins, sulfa drugs, and the like.[71]

Over the last thirty years, scores of studies have confirmed that this steady diet of antibiotics breeds highly resistant microflora in an animal's digestive tract and on its skin, as well as in the air, soil, and groundwater in and around livestock operations.[72] Analyses of supermarket meat and eggs show that at least some of this drug-resistant microflora also ends up shrink-wrapped with the meat we buy and trapped inside eggs before their shells form. From there, even the most meticulous cook will spread a few invisible contaminants around the kitchen and occasionally allow a few more to reach the dinner table in a less than thoroughly cooked burger, chop, or omelet.

Each year, salmonella and campylobacter, the most common causes of bacterial food poisoning, send an estimated 3 to 4 million Americans to the doctor or hospital for treatment, though many more suffer a miserable day or two close to a toilet before recovering on their own. In the most serious cases, when the bacteria spread beyond the intestines, effective antibiotics can spell the difference between full recovery and serious organ damage or death. But over the last two decades, increasingly multidrug-resistant strains of salmonella and campylobacter have emerged, first in animals and then in people.

"Salmonella is bad news. Drug-resistant salmonella is really bad news," says CDC epidemiologist Tom Chiller, the medical agency's liaison with the National Antibiotic Resistance Monitoring System, a collaboration with the U.S. Department of Agriculture and the veterinary branch of the FDA. The number of multidrug-resistant cases of severe salmonellosis shot up in the 1990s, Chiller reports, reaching a peak in 2000, when they accounted for around 40 percent of all cases. Since then the increases have concentrated in the most highly drug-resistant strains, those impervious to nine or more previously effective drugs. Mysteriously and dangerously, these superbugs tend to cause more-invasive infections, organ damage, and fatalities, even when doctors deploy effective antibiotics.[73]

"The biggest issue," says Chiller, "is that we're seeing increasing resistance to our most important antibiotics, to the two major classes of drugs that a physician is going to want to use with these infections." The first class comprises the fluoroquinolones, such as ciprofloxacin, or Cipro; the second, the cephalosporins. Unless they encounter resistance, Cipro and other "floxacins" are the most immediately effective treatments, by far, against severe salmonellosis and campylobacter infections. When treating children, however, doctors default to cephalosporins because of concerns about fluoroquinolone toxicity, especially in regard to a child's growing bones, ligaments, and nervous system.

"This is not the kind of resistance that a physician with a critically ill patient wants to see," says Chiller. Calls for government action increased in 2003 when monitoring showed that Cipro resistance had jumped to around 15 percent of all campylobacter bacteria contaminating chicken and eggs on supermarket shelves. That spurred the FDA to yank enrofloxacin, or "animal Cipro," from use on poultry farms in 2005. The move, the first time the U.S. government had withdrawn any antibiotic from the livestock market, marked a watershed in the on-again, off-again campaign against the U.S. livestock industry's controversial practice of using antibiotics from the same families of drugs as those actively used in human medicine.

Larger concerns center on the suspicion that the tonnage of antibiotics fed to livestock is fueling a pipeline of resistance genes from their microflora to ours. For every salmonella or campylobacter that sickens someone, we have to assume that hundreds to millions of normally harmless bacteria such as enterococcus and enterobacter are continually moving from farm to fork, says Chiller. "They don't have to make us sick in order to swap genes with other bacteria in our intestinal tract."

But while logic and circumstantial evidence support such a scenario, the livestock and pharmaceutical industries have long pointed out that no one has ever actually documented a bacterium from a cow, chicken, or pig passing off its resistance genes to a human pathogen outside the artificial confines of a test tube. "Who's to say it's not the other way around, that the tremendous amount of antimicrobial use in humans is actually driving up resistance in animals?" asks veterinarian Richard Carnevale, the Animal Health Institute's vice president of scientific affairs. Carnevale points to research out of Loma Linda University showing that meat eaters harbor no more drug-resistant microbes than do vegetarians (though the vegetarian volunteers in the study consumed both eggs and milk).[74] Others have found significant levels of drug resistance in the bacteria contaminating meat labeled "raised without antibiotics," he notes.[75] "You can even find resistant microbes in deer feces picked up from the middle of a forest."

A factor that researchers are only beginning to consider, one that may explain some of these apparent anomalies, is what happens to the tons of antibiotics that livestock excrete each year, Chiller says. Studies show, for example, that farm runoff spreads antibiotics into groundwater and streams.[76] It also defies logic, he adds, to suggest that the dramatic rises in resistant bacteria being documented on supermarket shelves are flowing from people to livestock and not the other way around. "So far as I know, nobody's feeding us to them," he says.

Several studies provide strong circumstantial evidence for livestock-to-human transfer of drug resistance. In the fall and winter of 1999–2000, for example, epidemiologists investigated an outbreak of highly resistant urinary tract infections among students at the University of California at Berkeley. DNA fingerprinting revealed that the UTIs stemmed from the same clone, or unique substrain, of the intestinal bacterium *E. coli*. That told the investigators that the *E. coli* had come from a common source. As few of the women either roomed together or even knew each other, they concluded that source to be contaminated food. When the investigators found the same clone causing UTIs in Michigan and Minnesota between 1996 and 1999, they realized that the food contamination was not isolated but likely came from the intestinal *E. coli* of livestock, nationally distributed in beef.[77]

Ironically, some of the strongest evidence of dangerous resistance genes moving from livestock to humans stems from a debacle created by early European efforts to restrict antibiotic use in food animals. As far back as the 1960s, British physicians were lobbying their government to investigate the role that livestock antibiotics played in causing the increasingly drug-resistant salmonella and campylobacter infections that were turning up in their clinics and hospitals. The result was the Swann Committee report of 1969, which showed a clear correlation between growth-promoting antibiotics and rising levels of resistant bacteria in the bodies of slaughtered animals, as well as a historic increase in drug-resistant salmonella infections in people.[78] In response, the British government banned its agricultural industry from using any class of antibiotic currently in use in human medicine for low-dose "nutritional" purposes. Much of western Europe followed suit, intending to preserve the effectiveness of medicinal antibiotics.

As a result, the livestock industry turned to an array of back-shelf drugs that had been deemed unsuitable for humans because of concerns about their toxicity or poor absorbability. Chief among them were the glycopeptides, the family of antibiotics that includes vancomycin. At the time, few imagined that, in the coming decades, this class of drugs would become the last hope of hospitals facing a rising epidemic of methicillin-resistant infections. To their great shock, when European doctors first resorted to vancomycin in the late 1980s, they found that resistance to it was already widespread among the intestinal bacteria of the general populace. Subsequent studies showed that millions of Europeans were walking around with VRE that carried vancomycin-resistance genes virtually identical to those found in the intestinal bacteria of chickens and pigs raised on avoparcin, another glycopeptide antibiotic and near chemical twin of vancomycin.

Though no eyewitness had observed the handoff of genes from livestock to people, the evidence was overwhelming. By contrast, in the United States and Canada, where farmers had not used avoparcin—given the abundance of other antibiotics available to them—VRE in people remained rare to unknown outside hospital patients. Of interest to microbiologists, the kinds of VRE inhabiting Europeans were clearly different from the kinds inhabiting their food animals. It was the resistance genes that proved the near-perfect match. In other words, the intestinal bacteria of animals were not infecting people; the livestock bugs were simply spreading their genes as they passed through a person's intestinal tract.

Denmark, the first European country to ban avoparcin in 1995, saw an immediate drop in vancomycin-resistant bacteria in its poultry—but the prevalence in its pigs didn't budge. The problem turned out to be cross-resistance. In the intestinal bacteria of pigs, the genes for vancomycin resistance sat adjacent to a gene conveying resistance to another antibiotic, tylosin. When the Danish government added tylosin to the growth-promoter ban, it produced the desired drop in vancomycin resistance in the bacteria contaminating pork products. Germany, Belgium, and the Netherlands followed the Danish ban in 1996. Over the next five years, clinical studies showed that the overall prevalence of VRE in the European population was halved, from around 12 to 6 percent; but it never returned to pre-avoparcin levels.[79]

In 1999 the European Union ordered a complete halt to the use of antibiotics as livestock growth promoters, a ban that was phased in over the following seven years. The trigger for the drastic move: In 1999 German researchers tested the effectiveness of Synercid, the promising new drug for VRE infections; they immediately encountered widespread resistance in patients and healthy volunteers alike. Like vancomycin, Synercid had been hammered out of a class of drugs that were little used in human medicine (the streptogramins) but were popular for livestock growth promotion.[80]

That same year, in the United States, CDC scientists detected Synercid resistance in nearly 90 percent of the enterococcal bacteria contaminating grocery-store chicken and 12 percent of the enterococcus isolated from healthy human volunteers.[81] Using Synercid in patients would quickly conflate that 12 percent to the levels seen in livestock, the researchers warned. True to predictions, in 2006 researchers at the Minnesota Department of Health reported they encountered Synercid resistance in 40 percent of the enteroccocus isolated from people who reported either handling raw poultry or eating a lot of chicken. By contrast, they found no such resistance in a comparison group of 65 vegetarians.[82]

Today Canadian and U.S. health and agricultural agencies are still struggling to come up with restrictions on agricultural antibiotics that somehow balance economics and public safety. Some have warned that the North American livestock industry—with its emphasis on large-scale factory farming and low profit margins—will have a tougher time adjusting to any ban on growth-promoting antibiotics than have the generally smaller-scale operations typical of Europe. In addition, industry groups have been quick to point out instances in which the European ban on low-dose antibiotics in the feed of healthy animals has led to increases in infections that require treatment at higher doses, potentially increasing the total amount of drugs given to an animal.

THE ANTIBIOTIC PARADOX

By their very nature, antibiotics sow the seeds of their own destruction, Stuart Levy warned in *The Antibiotic Paradox*, his 2001 manifesto on the dangers of antibiotic misuse and overuse. At the time of the book's publication, Levy had already become the media's iconic prophet of gloom and doom, his Groucho Marx mustache and features and his flamboyant bow ties making him particularly memorable on TV news exposés. And his warning has been borne out: drug-resistant bacterial infections now kill tens of thousands of Americans each year, this in a country where money is no object in the choice of a lifesaving antibiotic. Even though prudent use will prolong an antibiotic's effectiveness, Levy argues, these drugs are, at best, short-term solutions to the age-old problem of infectious disease.

PART V

FIGHTING SMARTER, NOT HARDER

A SHORT HISTORY OF MEDICINE

- 2000 B.C.E.—Here, eat this root.
- 1000 A.D.—That root is heathen. Here, say this prayer.
- 1850—That prayer is superstition. Here, drink this potion.
- 1920—That potion is snake oil. Here, swallow this pill.
- 1945—That pill is ineffective. Here, take this penicillin.
- 1955—Oops . . . bugs got resistant. Here, take this tetracycline.
- 1957–2007—42 more "oopses" . . . Here, take this more powerful antibiotic.
- 20??—The bugs have won! Here, eat this root.

—Anonymous

THE GOOD OLD DAYS?

At the height of the Civil War and barely a year into his service with the Union Army, twenty-seven-year-old Valentin Keller, a small, slender tailor from Ohio, received his medical discharge. "Unable to walk without the aid of crutches and then only with great pain," wrote the discharging physician, who also noted decreased breathing sounds, most likely "pleurisy," or fluid around the lungs. Like so many of his peers, Keller had been disabled not by bullet or bayonet, but by infectious disease. Crippled and in constant pain, he would die at age forty-one of "dropsy," or congestive heart failure, probably stemming from a childhood bout of typhoid or rheumatic fever.[1]

To explore the problems created by our war on germs is to risk suggesting that humanity was better off in the centuries before public sanitation and antibiotics disrupted our "natural" relationship with microbes. Today's antivaccine movement implies as much in its argument that, by depriving children of once-common infections like measles, mumps, and chicken pox, we're leaving them more prone to disease than their heartier ancestors. Even medical experts have questioned whether in keeping people alive longer, modern medicine has substituted quantity for quality of life—giving us not "golden years" but sadly debilitated ones. In the 1980s and 1990s, fears about the financial burden this would place on society spurred a slew of studies by economists.

Chief among these economists has been the team of Nobel laureate Robert Fogel and his protégée Dora Costa. They have spent the last decade mining our richest source of information on the health of nineteenth-century Americans: the medical records of the Union army and its Civil War veterans. Perhaps the saddest snapshot comes from the Civil War's induction exams, for they show that the recruitment-hungry Union Army reluctantly turned away tens of thousands of teenage boys and young men as too sickly and crippled to serve. Some of their debilities traced back to child labor in an age before occupational safety laws. But far more stemmed from childhood illnesses such as measles, typhoid, and rheumatic fever, which produced pervasive pleurisy, dropsy, and rheumatism (inflammatory arthritis).

Those who passed muster (neither the Union nor the Confederate army was picky; they accepted the half blind and the incontinent) encountered even more virulent disease in crowded battlefield trenches and encampments. Fifty percent of the Civil War's 620,000 fatalities resulted from disease, the worst stemming from rolling outbreaks of measles, diphtheria, typhoid fever, and strep infections. The toll on the survivors becomes clear in Costa's review of veteran medical exams: each fever that a soldier experienced during the war dramatically increased his risk of atherosclerosis, or hardening of the arteries, in middle age. Soldiers who had survived typhoid or rheumatoid fever went on to suffer high rates of valvular heart disease and arthritis. Tuberculosis predisposed them to chronic respiratory problems.

"Union Army veterans were already disabled by chronic conditions by age fifty," Costa says. In terms of overall health and mobility, she adds, "the fifty-year-old Civil War veteran of 1890 resembled his seventy-five-year-old descendants of today." Nearly half of Civil War veterans age fifty to sixty-five suffered painful arthritis, compared to 10 percent of men the same age today. Nearly a third had heart murmurs, and one in five had valvular heart disease, damaged lungs, or both—disorders that afflict fewer than 5 percent of their same-age descendants today.[2] Far from being "sadly debilitated," the health of the elderly in the United States has improved markedly in the twentieth century, Costa concludes.

Further insight comes from Costa's observation of an intriguing difference in survival rates between Union soldiers raised in urban areas and their country cousins. The city boys survived wartime epidemics in far greater numbers than did their rural counterparts. But as veterans the country boys survived to live, on average, several years longer. Both effects are the result of the greater burden of infection suffered by children in teeming nineteenth-century city slums: those lucky enough to survive entered the war at least partially immune to the germs running rampant in crowded wartime conditions; but their greater lifetime burden of infection took its toll in their later years. Several recent studies have confirmed this direct association; the more infections a person experiences, the greater the likelihood of arthritis, heart disease, stroke, and even cancer by middle age.[3] The link between the two: the inflammation that lingers long after the infection is gone.

Far from being universally "bad," inflammation simply represents an immune system that is set on high alert—armed and ready for the next invader. But like a jittery police force, these same inflammatory cells and chemicals have a tendency to shoot up the neighborhood— that is, healthy tissue. The resulting damage takes myriad forms, from the pain of inflammatory arthritis to the blood-vessel scarring of atherosclerosis.[4]

As for those who would call routine infections "natural," they mistakenly focus on the small fraction of human history that followed the rise of civilization some five thousand years ago. As mentioned in Part I, civilization's new dynamic of crowding and settlement fostered the rise of microbes that debilitated and killed their hosts. As for the quarter-million years that our species spent largely free of person-to-person contagion, we are now catching our last glimpse of this precivilized state, in the Amazon Basin's last nomadic Stone Age tribes.

Photographs and medical examinations of Amazonians such as the Nukak attest to their robust health—a state of grace that they lose within weeks of their emergence from the rain forest.[5] Anthropologists have expressed shock at how easily the Nukak fall ill, contracting serious infections not only from exposure to villagers but even from seemingly harmless microbes in the soil around permanent settlements. This extreme vulnerability to infection involves more than the Nukak's lack of previous exposure to the germs of civilization. Recent research suggests that in an environment where deadly infections remain rare, natural selection favors a mild, or tolerant, immune response—because it decreases a woman's risk of miscarriage up to sixteenfold.[6] A budding embryo, after all, is the ultimate "foreign" invader and demands enormous tolerance from the immune system if a pregnancy is to succeed.

By contrast, life in communities that are plagued by infection strongly favors the survival of babies who are genetically equipped for a brutally strong inflammatory response. The increased risk of miscarriage becomes a small price to pay for higher odds that a child will at least survive to reach reproductive age.[7] In this way some five thousand years of civilization and its attendant plagues have bred a high degree of aggressiveness into the "civilized" immune system, an inherent aggressiveness that is intensified by infections small and large.

In a crude way, public sanitation, antibiotics, and childhood vaccinations have partially restored humanity's precivilized state of health, for they have drastically reduced the typical person's lifetime burden of disease-related inflammation. But they have done so without changing the genetic underpinnings of our aggression-bent immune systems. And in the case of sanitation and antibiotics, they have done so by crudely sweeping away life's harmless, immune-calming bugs along with the disease-causing, inflammatory ones. The result appears to be a redirection of immune aggressiveness to the "imagined" threats in allergens, and perhaps the body's own healthy cells.

Today's challenge, then, is not to abandon sanitation and antibiotics but to preserve their effectiveness and rectify their side effects. This challenge has proven particularly hard in regard to antibiotics, given that resistance is an inescapable consequence of their use. In fact, many pharmaceutical companies, repeatedly stung by the development of resistance to their newest products, have chosen to give up the fight. In response, the physicians and scientists of the Infectious Diseases Society of America (IDSA) have been lobbying the U.S. government to provide the generous research grants and tax breaks needed to bring pharmaceutical companies back into the questionably profitable business of antibiotic discovery. "Who wants to invest millions of dollars to develop a new drug whose widespread use is guaranteed to render it ineffective?" asks IDSA president Martin Blaser. It's also fair to ask whether a reinvigoration of antibiotic development will produce more of the same approach that led to our current crisis in drug resistance. Have we learned enough from fifty years of squandering the effectiveness of our antibiotics to rethink how we develop and deploy these miracle drugs?

PRESERVING ANTIBIOTICS: LESS IS MORE

While doctors and patients clamor for newer and more powerful antibiotics, it's clear that we're still not using our existing drugs responsibly. We know, for example, that the less we use them, the longer they will remain effective. But two decades of hectoring physicians to curb "antibiotic abuse" has produced limited results. A key point of contention is the stubbornly persistent practice of keeping healthy but infection-prone patients on prophylactic, or preventive, antibiotics, for months to years on end—something known to breed drug resistance into patients' microflora and even that of their housemates.[8] In 1999 the American Academy of Pediatrics began actively discouraging its members from the habit of giving ear-infection-prone babies and preschoolers daily amoxicillin. Yet it remains common for dermatologists to prescribe long-term antibiotics to ease teenagers through their acne-prone years, and gynecologists often do the same with women predisposed to urinary tract infections.[9]

Add to this the growing popularity of using antibiotics for their anti-inflammatory effects in treating nonbacterial diseases such as asthma, rheumatoid arthritis, and even obsessive-compulsive disorder.[10] The popularity of this approach took a jump in the 1990s with the discovery that several classes of antibiotics, including tetracyclines and macrolides, somehow dampen the inflammatory activity of the immune system's soldier cells.[11] But because antibiotics don't cure the underlying cause of these disorders, their effectiveness depends on continual, even lifelong use.

Cardiologists, in turn, have expressed great interest in using antibiotics to treat atherosclerosis, or clogging of the arteries, which we now know to be an inflammatory disease. The inflammation may stem from trace amounts of bacteria that stray from the mouth and respiratory tract and end up, in dormant form, in the lining of our blood vessels.[12] No one knows why some people's immune systems ignore these otherwise harmless bacteria while others respond with artery-clogging inflammation. Regardless, many cardiologists began putting their patients on antibiotics in the 1990s, with the hope of curing heart disease.[13] The practice largely stopped in 2005, when two large studies showed that the drugs neither eliminated the bacteria nor reduced the risk of heart attack.[14] However, these studies left open the possibility that longer courses of more powerful antibiotics might produce the desired effect. Should such an effective regimen be found—and several companies are betting their profit margins on finding one—the result could be tens of millions of middle-aged men and women on antibiotics in the United States alone.[15]

Scientists concerned with drug resistance warn that such increasingly widespread use will have dire consequences. But most physicians bristle at the idea of being limited in what they can prescribe to whom. In the short term, relieving one patient's socially crippling acne or reducing another's risk of heart attack tends to trump the seemingly distant threat of fostering dangerous drug resistance—at least until a doctor begins losing patients to the kind of unstoppable bacterial infections that now kill more than fourteen thousand Americans a year.[16]

Doctors may well feel unfairly blamed for antibiotic "abuse" when their prescriptions account for less than half the tonnage sold each year. "Even if we drastically reduce our prescriptions, given all the antibiotics in animal feed, we'll probably still have resistance," asserts Jim King, president of the American Academy of Family Physicians and a private practitioner in Selmer, Tennessee. Supporting his argument are recent studies showing that the drug-resistant bacteria bred in large-scale livestock operations not only reach us in meat and eggs, but travel from factory farms in wastewater and rain runoff to end up in our rivers and reservoirs.[17] Nonetheless, it's clear that antibiotics affect us—that is, our resident microbes—most directly when we swallow or otherwise infuse them into our bodies as concentrated drugs. In this regard, experts have itemized several powerful ways that physicians can slash their patients' exposure:

- **Eliminate unnecessary prescriptions.** Doctors today are definitely less likely than they were a decade ago to prescribe antibiotics when none are needed—that is, for conditions such as viral infections that don't respond to the drugs. But unnecessary prescriptions still account for around one-third of all the antibiotics we take.[18] In survey after survey, physicians claim that the main reason they prescribe antibiotics unnecessarily is "patient insistence."[19] Behind this strange implication—that the sick are strong-arming their doctors—is the admission that physicians find it faster to scribble a prescription than to explain to a patient that antibiotics work only against living organisms such as bacteria, not against viruses—the cause of most respiratory infections, including colds and flu. "Sometimes it's just easier to write the prescription the patient wants," admits King. "Easier, though obviously not right."

- **Get patients off antibiotics faster.** In 2006 a team of Amsterdam doctors discovered that the standard prescription for treating bacterial pneumonia—ten days on amoxicillin—is about seven days too long. Three days works just as effectively in clearing the infection, they found, dispelling the myth that shorter courses risk leaving the hardiest bugs behind to trigger resistant rebound infections.[20] What's more surprising, perhaps, is that until recently no one had challenged the standard length of antibiotic treatment for this and many other common bacterial infections. "A lot of this is simply passed on by tradition rather than evidence," says Case Western's Curtis Donskey. The good news is that the Dutch study has prompted at least a dozen similar reassessments. "In all of the studies reported so far," says Donskey, "the patients did just as well with the shorter courses of antibiotic." It may take years for the new practice guidelines to become widely accepted, Donskey adds, cautioning that it's still unwise for patients to take such matters into their own hands: some deep-seated infections can rebound in more severe form when inadequately treated.

- **Use less-disruptive drugs.** "An antibiotic prescribed to treat an infection penetrates all body compartments, selecting for bacteria with resistance and changing our microecology, perhaps permanently," warns IDSA's Blaser, also chief of medicine at New York University Medical Center. The worst offenders in this regard are the so-called broad-spectrum antibiotics, long touted by pharmaceutical companies for their ability to stop almost any kind of bacterial infection. More sparing of the microflora—and so less likely to breed resistance—are narrower-spectrum drugs such as the older penicillins and erythromycins (macrolides), which can take out a bug such as *Streptococcus pyogenes*, the bacterium causing strep throat, while leaving bystander bacteria relatively unscathed. Unfortunately, doctors today are much more likely than they were ten to fifteen years ago to reach for the big-gun, whatever-might-ail-you drugs, due to their convenience and the heavy promotion by drug companies.[21] This disturbing trend, resistance experts argue, has more than canceled out progress in getting doctors to write fewer prescriptions overall.

The crux of the problem, adds King, is that identifying the specific bacterium that causes an infection and determining its drug susceptibilities take time and effort. "In a busy private practice, there's no way we're going to run cultures to identify every organism causing an infection," he explains. Culturing an organism from a throat swab or urine sample, for example, typically takes twenty-four to forty-eight hours; testing for drug susceptibility can add another day.

"That's not going to happen," says King. "As a physician, I'm going to pick an antibiotic that, in my experience, gets the job done." Ideally, King adds, the responsible physician will prescribe the narrowest-spectrum drug that's likely to clear a given type of infection, then switch to a bigger gun if the patient doesn't feel better within a day or two. "But sometimes it's just easier to prescribe the broad-spectrum and not have to worry about follow-up," he confesses.

Such "shoot first, aim later" habits become more justifiable in hospitals, where lives depend on quick use of an effective antibiotic. But the critical-care use of broad-spectrum antibiotics comes with a far greater risk—disrupting the body's normal microflora in a way that invites invasion by hospital superbugs. For all these reasons, resistance experts agree that the development of rapid pinpoint diagnostics will prove to be the key to reforming prescribing practices.

HOMING IN ON THE ENEMY

In the fall of 2005, University of Florida obstetrician Rodney Edwards gave his delivery room nurses a new toy: a gene-amplifying gadget that was able to detect DNA fingerprints faster than anything in the most tricked-out forensics lab. "They loved it," says Edwards of the enthusiasm the nurses brought to their task, which was to rapidly identify the presence of *Streptococcus agalactiae*, or group B strep, in the birth canals of women in labor. The bug is part of the normal intestinal microflora, and in 20 to 40 percent of women part of the vaginal flora as well. It is harmless, except to an emerging newborn. Around one in a hundred GBS-infected babies develops a life-threatening infection of the bloodstream or cerebrospinal fluid. So obstetricians routinely test women for the bacterium between their thirty-fifth and thirty-seventh weeks of pregnancy, sending a vaginal swab off to the laboratory for a three-day culture and placing those who test positive on antibiotics during labor and delivery.

The problem is, many women end up in the delivery room without GBS results, often due to premature labor, delayed culture results, lost paperwork, or a lack of basic prenatal care. Consequently, delivering physicians must needlessly place millions of laboring women—and by extension their newborns—on antibiotics each year to guard against a bacterium that *most* women don't carry. And that puts both women and babies at risk of picking up drug-resistant hospital infections.

The gadget that Edwards had his nurses test in 2005 and 2006 has the power to change that situation. The nurse gently passes a swab across the outside of a woman's vagina; inserts the specimen end of the swab into a small cartridge; squirts a premeasured tube of "Reagent 1" into a hole marked "Port 1," and "Reagent 2" into "Port 2"; then snaps the cassette into one of four docking bays on a tabletop gene amplifier. Total time to collect and process specimen: two minutes. Time to results: one hour.

"I do think this is the way of the future, not that it's the way to screen everyone," says Edwards, who was testing the experimental $40,000 DNA analyzer, called GeneXpert, for Cepheid Diagnostics. While in theory the device could amplify and detect the signature DNA of any number of microbes—and their drug-resistance genes—at present only two such "real-time PCR," or rapid gene amplification, tests have FDA approval. The one Edwards tested was approved in mid-2006. The second, a two-hour test for staph, takes rapid diagnostics a vital step further by adding drug susceptibility to the results: it not only tells doctors whether a patient has *Staphylococcus aureus*, it also reveals whether it is methicillin resistant, as more than half of all staph bugs are in many major medical centers. When doctors have to make this determination the old-fashioned way (by growing a bacterial culture much as Pasteur would have done a century ago and then testing it against various antibiotics), getting the result can take up to three days—three days during which doctors must hedge all bets by using the most powerful drugs in their arsenal.

Many medical experts call devices such as the GeneXpert a "breakthrough," not so much for their ability to perform DNA fingerprinting as for their ease of use.[22] Around a dozen gene tests for bacterial infections have won FDA approval in recent years. But until now all took some sophisticated laboratory finagling to perform and a half day or more to produce results. "If we're going to get more doctors, especially community doctors, to use diagnostics to guide their prescribing," says Roberta Carey, a CDC specialist in diagnostic microbiology, "we're going to have to make it so foolproof that even on your worst day, you can answer a phone call in the middle of performing the test and it will still work. And it has to be quick, or doctors won't bother."

As a family physician, King agrees: "I want to give my patient a prescription now, not in a day or two." For the time being, DNA detection devices such as Cepheid's GeneXpert remain too expensive for use outside the hospital, and general physicians have only one practical in-office diagnostic at their disposal: the RapidStrep test that detects the presence of *Streptococcus pyogenes* from a throat swab in around ten minutes. The test detects not DNA but large molecular markers on the slippery capsule that coats this bacterium. Even this one test, when used conscientiously, could make a significant dent both in the over-prescribing of antibiotics and, by extension, in rates of drug resistance in a doctor's community.

Just how big a dent can be seen from the results of a recent real-life experiment in which French researchers targeted several towns with intensive information campaigns aimed at educating kindergartners, their parents, and doctors about prudent antibiotic use—specifically, about not using these drugs for throat infections unless a rapid-antigen test confirmed the presence of *Strep. pyogenes*, the only bacterium commonly causing sore throats (the vast majority of which are caused by viruses).

By the end of the four-month education campaign, antibiotic use in the communities had dropped by nearly 20 percent. Moreover, the researchers found an abrupt drop in dangerous drug resistance in another throat bacterium, *Streptococcus pneumoniae*, a major cause of pneumonia and meningitis as well as chronic ear infections. In the communities where faithful use of the strep test reduced antibiotic prescriptions, the percentage of kindergartners carrying drug-resistant strains of *Strep. pneumoniae* dropped from more than half to just over a third.[23]

Unfortunately, doctors have only a small handful of rapid-antigen tests like the ten-minute strep screen, as their effectiveness tends to hinge on a bacterium wearing a fairly large and complex antigen on its surface. Most bacteria lack such molecular billboards.

A bigger problem with such tests — be they for antigens or for signature genes — is that they don't tell the doctor what *else* might be causing problems, says Michael Dunne, head of clinical development for the pharmaceutical giant Pfizer in New London, Connecticut. "When you ignore the other bugs, you risk treating to the test, while the patient dies of something else," he explains. A wound infected with staph, for example, might be co-infected with *Streptococcus pyogenes* or *Pseudomonas aeruginosa*, either of which is capable of producing a deadly blood infection if ignored.

While Dunne says he would like to take Pfizer's antibiotic research in a new direction, away from broad-spectrum drugs and toward the kind of "sniper bullets" that can take out a specific species, he adds that this can't happen unless doctors have pinpoint diagnostics that can scan for a wide range of microbes and their drug susceptibilities.

In a fluky convergence of names and interests, another Michael Dunne, a molecular microbiologist at Washington University, described just such a device in a 2003 issue of the *Journal of Clinical Microbiology*.[24] In his admittedly futuristic essay, Dunne Two describes a typical morning in the practice of Jeffrey Lane, who is three years out of medical school in the year 2025. Lane is taking a throat swab from a sixteen-year-old who is complaining of severe pain on swallowing, headache, and nausea. After swiping the swab across the young man's tonsils, Lane inserts its tip into one end of his "MyCrobe" infectious-disease diagnostic unit — in appearance not so unlike the gadget wielded by *Star Trek*'s fictional Dr. Leonard "Bones" McCoy. Into the handle of the unit Lane slides one of five MyCrobe cassettes at his disposal, this one for upper respiratory infections. The cassette's inner chamber is impregnated with gene probes specific for more than 150 kinds of bacteria, viruses, and fungi known to cause such infections, as well as for several thousand bacterial genes for different types of virulence and drug resistance. A second chamber in the handheld device contains molecular probes for proteins and other molecules that determine whether any of these resistance and virulence genes are actively producing toxins and drug-deactivating enzymes, a sign that a bacterium isn't just a bystander but is actively causing disease. Before the patient leaves the office fifteen minutes later, the device has registered more than a dozen DNA and RNA sequences specific for *Streptococcus pyogenes* and determined it to be resistant to penicillins, methicillins, cephalosporins, macrolides, and spectogramins. This information guides Lane's prescription of an old-fashioned beta-lactam-beta-lactamase inhibitor.

Since Dunne's essay was published, researchers have tested three prototypes of such a diagnostic tool, and one is close to hitting the market. The threat of bioterrorism produced the first, a DNA microarray developed by scientists at Lawrence Livermore National Laboratory, in Livermore, California. It simultaneously screened for the presence of eighteen different deadly microbes and viruses, including eleven kinds of bacteria from *Bacillus anthracis* (anthrax) to *Yersinia pestis* (plague) in six hours.[25] The microarray registered the presence of each target bacterium's 16S rRNA gene (the species-specific signature that microbiologists use to sort through mixed samples of unknown bacteria; see Part II). "This was impressive for 2002," says Lawrence Livermore bioinformatics expert Tom Slezak. "But now we know that 16S rRNA analyses only go so far. In some cases they can't reliably identify the species." An even greater barrier to clinical use, he adds, was expense—each run of the microarray cost several hundred dollars.[26] Yet the prototype remains an important one; the price of DNA microarrays tumbles year by year, and better gene targets have already been identified and are waiting to be incorporated into next-generation technology.

The second stab at producing something resembling Dunne's "MyCrobe" came out of the University of Medicine and Dentistry of New Jersey in 2004, with the aim of detecting the most drug-resistant and deadly strains of staph. The device screened for the presence of six bacterial genes in a patient sample. The first three confirmed the presence of *Staphylococcus aureus*; the others signaled methicillin resistance, vancomycin resistance, and the toxin Panton-Valentine leukocidin. In under three hours the assay flashed the results via an array of multicolored lights. The problem was, the blinking lights didn't tell the researchers whether the resistance genes they were detecting were inside *Staph. aureus* or in a "bystander" bacterium also present in the sample. Methicillin-resistance genes, for instance, can turn up in the skin bug *Staphylococcus epidermidis,* and vancomycin resistance is common to hospital strains of the intestinal bacterium enterococcus.[27]

In 2005 the San Diego biotech company GeneOhm began testing a second-generation diagnostic for MRSA with an elegant solution to the "which gene is in which bug?" problem. Its two-hour MRSA assay contains two linked gene probes, one of which latches onto a stretch of DNA associated with methicillin resistance, the other a species-specific gene for *Staph. aureus.*[28] Since the double-ended probe can reach only so far, it has to find both targets on the same chromosome (i.e., in the same bug) to register a positive result.[29] After a successful run through clinical trials, the device looked like a good prospect for FDA approval in 2007.[30]

"I believe we're on the verge of a breakthrough in diagnostics," says Pfizer's Dunne, "one that could lead to finding drugs that work not only on the species but the strain and perhaps the specific organism. The idea of blowing away all the 'good' strep in a patient's throat when you're trying to eradicate *Strep. pyogenes* would become a thing of the past."

DRUGS WITH ON-OFF SWITCHES

Meanwhile, there may be ways to let doctors have their broad-spectrum antibiotics while at least partially curbing the drugs' harm. The Finnish pharmaceutical company Ipsat has developed an "intestinal protection" enzyme that destroys leftover antibiotics before the gallbladder secretes them into the colon—where the vast majority of our native bacteria reside. Early trials in both animals and patients show that the deactivating enzyme doesn't reduce the drugs' effectiveness where they're needed—in body tissues—but does prevent the usual disruption and drug resistance in intestinal bacteria.[31] So far, the company has deactivating enzymes for penicillins and the wide-spectrum cephalosporins and carbapenems, which account for nearly half of all antibiotic prescriptions. However, the enzyme works only with antibiotics that are administered by injection or intravenous line—which is perfect for hospital patients but of little help for those taking their prescriptions by mouth.

For oral antibiotics, which get absorbed through the intestinal tract, the challenge is the opposite: to keep the drugs inactive until *after* they get absorbed. Drugs that can pull off this trick are known as "prodrugs," and familiar nonantibiotic examples include Levodopa, a Parkinson's drug that begins working only after it crosses into the brain, and cancer chemotherapies that become toxic only after entering tumors. Some of the first prodrug antibiotics came about in the early 1990s, when bio-chemists began tinkering with the powerful new cephalosporin drugs to improve their absorption so they could be taken in pill form instead of injected. The chemists found they could greatly improve absorption by attaching a small compound (an ester) to a piece of the larger cephalosporin molecule. Further tinkering produced esters that conveniently fell away again as soon as the drug passed into the intestinal tissue, so as not to interfere with its action.[32]

The fortuitous side effect was a group of drugs that did not produce intestinal "upset"—that is, the kind of diarrhea that antibiotics provoke when they raze our digestive bacteria. Some prodrug antibiotics turned out to have the additional advantage of being excreted primarily in urine rather than in intestinal bile, so they bypassed the intestinal tract on the way out as well as the way in. The greater concentration in the urine also improves their effectiveness against urinary tract infections.[33]

In Europe, where prodrug antibiotics have become popular, studies have confirmed that the prodrugs pivmecillinam and bacampicillin have little effect on the resident bacteria of the mouth, throat, and digestive tract.[34] Pivmecillinam is particularly popular in Sweden, Denmark, and Norway, where it is the first choice for treating urinary tract infections, owing to its low rate of drug resistance (a world-wide problem among women afflicted with chronic UTIs). But prodrug antibiotics have yet to be introduced in the United States, where few physicians have heard of them.

SILENCING RESISTANCE

While minimizing an antibiotic's collateral damage can slow the further rise of drug resistance, it won't necessarily reverse the already dangerous levels we face today. Gone is the early hope that bacteria would jettison their resistance genes as soon as they no longer needed them. We now know that drug-resistant bacteria can compete quite well with their nonresistant kin, even in the bodies of those who haven't touched antibiotics in months or years.

Though it's naïve to hope we might ever restore resistance to the trace levels seen before penicillin, we may someday have the technology to selectively knock out bacterial resistance on a gene-by-gene, infection-by-infection basis. One promising avenue of research involves plasmid "emetics"—chemicals and other treatments that goad bacteria into disgorging these circlets of swappable DNA, which often come fully loaded with resistance genes. Efforts to cure multidrug-resistant bacteria of their plasmids go back to the 1970s, when microbiologists were just beginning to grasp what these "resistance factors" were. In the laboratory, researchers found they could sometimes purge a bacterial colony of its drug resistance by subjecting it to noxious chemicals such as acridine dyes, caustic detergents, or the carcinogen ethidium bromide.

At the time, such toxic chemicals seemed far from acceptable for treating patients. But the serious side effects of some of today's last-resort antibiotics have prompted scientists to give plasmid-purging chemicals a second look.[35] The occasional seizure caused by the antibiotic imipenem, for example, is now considered a "reasonable" side effect in the treatment of vancomycin-resistant enterococcus or multidrug-resistant pseudomonas and acinetobacter. Unfortunately, no one "curing agent" works on all or even most kinds of plasmids, and many bacteria tote around several.

A safer, more effective approach may come out of efforts to understand how bacteria pass their plasmid collections to their daughter cells. Like all cells, bacteria multiply by dividing, though in a much simpler fashion than do the larger, more complicated cells of plants and animals. Plasmids, it turns out, carry their own genetic machinery for independent replication and inheritance. Moreover, they must tightly control the process to ensure that there are always enough plasmids to get doled out to the next generation, but not so many that they become too numerous inside any one bacterium.

To Paul Hergenrother, a lanky chemistry wunderkind at the University of Illinois, the complexity of the plasmid-replication process provides scientists with ample opportunities to muck it up—either by tricking plasmids into "thinking" there are too many of them (and so shutting down replication) or by scrambling their ability to sort copies into daughter cells.[36] In 2004 Hergenrother's research group accomplished the former, using a small druglike molecule, apramycin, that mimicked a signal that said "too many plasmids." Inside E. coli cells bathed with the molecule, resistance plasmids stopped replicating—so that with each generation, the daughter cells had fewer plasmids to inherit. After some 250 generations (E. coli replicate, on average, every twenty minutes), Hergenrother's colonies became virtually plasmid-free—and fully susceptible to antibiotics that had not fazed them three days earlier.[37] Clearly, doctors would need a faster "plasmid purger" if they were going to use it in conjunction with antibiotics. But Hergenrother's success represents a vital step in that direction.

Another mode of attack involves finding the genes that some bacteria use to turn on and off their plasmid-borne resistance genes. This sort of "inducible resistance" is a maddening problem in treating certain types of bacterial infections, because a bug can test susceptible to a particular drug in the laboratory only to switch on its resistance genes during the patient's treatment. But such on-off switches present tantalizing new targets for resistance-reversing drugs. At England's University of Bristol, for example, molecular biologist Virve Enne has been searching for the switch that normally suppresses the activity of an entire E. coli resistance plasmid.[38] When switched on, this particular plasmid renders E. coli impervious to tetracycline, ampicillin, streptomycin, and sulfamethoxazole—four of the most widely used antibiotics in medicine. By early 2007 Enne had narrowed the "controller switch" down to one of several groups of genes. Once she has it isolated, she must then decipher what sort of signals switch it on and how she might block those messages.

Of course, plasmids are not the only repository of resistance genes. Some of these genes arrive in a bacterium aboard a plasmid, only to jump to the bacterium's main chromosome. Others arise by mutation or get inserted into the chromosome by bacteriophages (bacteria-infecting viruses) or conjugative transposons (a kind of "jumping gene"). The great hope in these situations lies in one or more methods of direct gene silencing, or antisense gene technology. These techniques involve artificially constructing snippets of reverse-order DNA that block the action of their mirror-image genes. The most familiar application of this technology was the ill-fated Flavr Savr tomato of the early 1990s. Genetic engineers created the slow-ripening tomato by inserting antisense genes that partially blocked a ripening hormone—allowing farmers to pick the tomatoes when they were already ripe (and more flavorful) without worrying that they would rot on the way to market.[39] The modest increase in flavor was not enough to surmount consumer discomfort with genetically engineered "Frankenfruit," but it was a great demonstration of the power of antisense science.

Around the same time that the Flavr Savr made its brief debut, Stuart Levy's team at Tufts began constructing an antisense gene to block the multiple-antibiotic-resistance (mar) gene found in a host of closely related pathogens, including salmonella (food poisoning), shigella (bacterial dysentery), yersinia (plague), and a dozen-odd hospital pathogens such as enterobacters, citrobacters, serratia, and klebsiella. Mar is a kind of master switch for an entire suite of resistance genes that produce immunity to dozens of antibiotics, from old standards like tetracycline and ampicillin to new hopefuls such as norfloxacin. Levy and his colleagues put their antisense-mar gene into plasmids and induced E. coli cells to pick them up with a one-minute blast of heat (108°F or 42°C) or pulse of electricity.[40] It was not a practical method for treating bacteria causing an infection, to be sure, but it opened a promising new avenue of research.

Since Levy's demonstration, other scientists have accomplished similar feats, goading test-tube bacteria to pick up antisense molecules that reversed resistance to vancomycin, in one experiment,[41] and to amikacin in another.[42] (A rather toxic drug, amikacin is most often used for severe hospital infections that are resistant to all else.) As with plasmid purging, the future success of these efforts hinges on finding practical ways of making bacteria cooperate inside patients. On the hopeful side, microbiologists are exploring several methods for slipping antisense genes into bacterial cells. For example, they might be packaged into phages or encased inside fat particles that pass easily through a bacterial cell wall.

FARMING OUT RESISTANCE

While the most direct cause of antibiotic resistance is prescription use, evidence continues to grow that resistant bacteria and their dangerous genes are reaching us via the meat, eggs, and polluted runoff coming out of livestock operations. A growing coalition of North American scientists, physicians, and consumers continues to urge the United States and Canada to end the practice of adding antibiotics to livestock feed to speed animal growth. The European Union completed such a ban in 2005. But even a complete halt to the use of antibiotics for growth promotion would leave us with the heavy tonnage that is administered to herds and flocks to treat and prevent the spread of suspected infections. And as lobbyists with the livestock-pharmaceutical industry have rightly pointed out, in some instances Europe's ban on growth-promoting antibiotics has actually led to increased sickness and therapeutic use.[43]

The solution? In advising Canadian regulatory agencies, University of Guelph veterinarian and food safety expert Scott McEwen urges a win-win compromise. "For us, progress has to come with a more nuanced approach," McEwen says. As an adviser to Health Canada, the Canadian government's equivalent to the U.S. Food and Drug Administration, McEwen has long straddled the fence between the industry lobby and consumer advocacy. "For me," he says, "the take-home message from Europe has been that there are lots of situations where we could dramatically reduce microbial resistance in animals without having much if any negative impact on a farmer's productivity." For instance, McEwen points out that withdrawing antibiotics from animals closer to maturity and slaughter might give their bodies a chance to re-populate with less-dangerous bacteria without any significant change in either growth rate or susceptibility to disease. By contrast, when farmers first remove piglets from their mothers and mix them together in a kind of vast piggy day care, respiratory and gastrointestinal infections tend to run rampant without some kind of preventive course of antibiotics fed to the whole lot.

At the same time, McEwen points out that ionophores—which account for as much as half of the "nutritional" antibiotics used in North American livestock—have no known or potential use in human medicine; nor have they been associated with cross-resistance to important antibiotics. "Why not use them?" he says, mystified at their banishment from European farms. "But if anyone tries to say we should use cephalosporins or fluoroquinolones as growth promoters, that's a no-brainer. We've seen that resistance to these classes develops quickly, with clearly deleterious effects on human health." The best compromise to date, says McEwen, may be the policy adopted by the American fast-food giant McDonald's in 2003. Its new requirements, applicable to all its meat suppliers worldwide, prohibit the use of antibiotics belonging to any class approved for human medicine—whether in active use or not—for solely growth-promotion purposes.[44]

As in human medicine, restricting the veterinary use of antibiotics—whether for disease prevention or for treatment—will be far trickier, McEwen concludes. Clearly, no one wants to leave sick animals untreated, whether because of empathy for their suffering or because of an aversion to eating diseased meat. "Then you get into proposals to restrict certain classes of drugs to use in people and others to use in animals," he says. "Except that doesn't work well. Just witness the example of avoparcin and vancomycin." Continual monitoring, he concludes, has to be a large part of any solution: that would involve maintaining databases on what antibiotics veterinarians and farmers are giving to livestock, collecting information on what kind of resistance genes are showing up in the bacteria that are contaminating meat and eggs, and analyzing the clinically important drug resistance that doctors encounter in their patients. "To date, the Danish have the gold standard for monitoring," he says, "with a highly regulated reporting system where nothing goes into a pig or other animal except by prescription, and then this information is collated for the entire nation." The cost and practicality of operating such a system in Canada and the United States—with their larger livestock industries and their hodge-podge of prescription and nonprescription livestock drugs—remains prohibitive, admits McEwen. "Maybe someday," he says. Till then the Canadian government has begun a prototype monitoring system that focuses on a network of sentinel farms and veterinary clinics that may provide an early warning of resistance on its way to the dinner plate.

We might make greater progress by shifting wholesale back to more traditional styles of farming, McEwen adds. Studies consistently show that animals raised on small "family" farms need less antibiotics and carry fewer drug-resistant bacteria. But such old-fashioned methods cannot match factory-style farming's cost-cutting economy of scale. "Few consumers would be willing to pay the added price," McEwen admits. Ultimately, the solution may lie in finding effective *alternatives* to antibiotics, he adds. In this regard, agriculture may benefit from novel therapies that are now being explored in human medicine.

BEYOND ANTIBIOTICS: NEW WAYS TO KILL

As snarls of chemicals go, a bacteriophage does a damn good imitation of a living organism. Most phages, when glimpsed through an electron microscope, resemble gangly-headed spiders, or perhaps lunar landers. When a phage's jointed "legs" contract, they bring the virus's "tail" into contact with a bacterium's cell wall, through which it injects the genes that will transform the cell into a phage-making factory. By comparison, the viruses that infect plants and animals resemble mere spitballs of protein-coated genes that stick to cells and get mistakenly absorbed.

The phages' greater sophistication is unsurprising, considering that their arms race with bacteria predates the arrival of all other life on Earth, perhaps by several billion years. This long evolution may also explain why phages turn out to be the world's most discriminating viruses. Many infect not just a particular species of bacterium but a few select strains. Before the advent of genetics, this discrimination gave microbiologists a handy way to identify and distinguish bacterial strains.

The 1950s baby-killer *Staph. aureus* 80/81, for example, got its designation by its susceptibility to phages 80 and 81 in a panel of more than a hundred.

Today phages are best known from such headlines as *Wired* magazine's "How Ravenous Soviet Viruses Will Save the World" (subhead: "They eat drug-resistant bacteria for breakfast").[45] The romantic history of phage discovery and use has inspired scores of magazine articles, TV documentaries, and books, from Peter Radetsky's "The Good Virus" in a 1996 issue of *Discover* magazine to Thomas Hausler's 2006 *Viruses vs. Superbugs*.[46]

The story of their medical use begins in 1916, when the French-Canadian microbiologist Félix d'Herelle first isolated bacteriophages in stool and sewage and went on to show that they grew only inside bacteria, which they effectively killed in the process. That same year, d'Herelle developed a phage cure for shigellosis, or bacterial dysentery. Largely dismissed in the West, d'Herelle left Yale University in 1933 to help one of his protégés, the Soviet microbiologist George Eliava, establish a bacteriophage research institute in Tbilisi, Georgia, with the generous support of Joseph Stalin. Though Eliava fell from favor and ended up in front of a KGB firing squad, phage therapy thrived behind the Iron Curtain, where it provided a cheap alternative to antibiotics. Even today Russians and Georgians can buy over-the-counter phage remedies for upset stomach, urinary tract infections, and a variety of other common maladies. And Tbilisi's Eliava Institute remains the world center of phage therapy. But the Soviets never poured much money into scientific studies of phage treatment, and the USSR's collapse in 1991 left Georgia's Eliava Institute destitute. So the rigorous proof that phage therapy actually works has been meager at best.

The credit for reintroducing phage therapy to the West goes largely to Alexander "Sandro" Sulakvelidze, a Georgian molecular biologist who came to the University of Maryland School of Medicine on a research fellowship in 1993. At the time, Sulakvelidze's supervisor, Glenn Morris, was struggling to control escalating outbreaks of vancomycin-resistant enterococcus at the university's hospitals in Baltimore. "Why don't you try phages?" Sulakvelidze suggested.[47]

As a treatment, phages sounded ideal: narrowly targeted, they promised to be sparing of the body's normal microflora and utterly harmless to its human cells. Moreover, any sewage-contaminated waterway offered an endless supply for screening out the ideal phage or phages to treat any given infection. By 1996 Morris and Sulakvelidze were gathering phage from the Baltimore harbor for laboratory experiments and presenting their test-tube findings at medical conferences.

To say the least, their ideas sparked interest. By 2002 more than two dozen start-up biotech companies were racing to bring phage therapies to clinical trials, in a frenzy matched only by the media coverage of what had been dubbed "Stalin's Forgotten Cure."[48] An almost cultlike following developed among patients with untreatable bacterial infections, spurring yet more stories in the media about desperate journeys to Tbilisi for cures unavailable in the West. In 2003 the Newark *Star-Ledger* featured the story of Kevin Smeallie, a thirty-seven-year-old American who was plagued by excruciating bacterial sinus infections resistant to all antibiotics. Smeallie's Internet search for "alternatives to antibiotics" had led him to the Eliava Institute. There he underwent surgical implantation of phage-impregnated "bioderm" strips in an operating room where, a month earlier, the single overhead light had fallen out of the ceiling during surgery. "Only the quick reflexes of one of the attending doctors prevented a catastrophe," wrote *Star-Ledger* reporter Amy Ellis Nutt. "He then held the light over the patient for the remainder of the operation." Smeallie enjoyed a one-month reprieve from his pain until, back in the States, his infection returned.[49] The article concluded with the American engineer on a new antibiotic and planning a return trip to Tbilisi. (He has since been diagnosed with an immune deficiency that leaves him prone to chronic infections.[50])

Smeallie is far from alone in his stubbornly persistent faith in phage therapy. But over a decade after phage mania began in the West, popular articles about the treatment still far outnumber anything in the scientific literature. Today the once-crowded field of companies pursuing phage research has dwindled to five or six, and none have reached patient trials. In late 2006 one company—Morris and Sulakvelidze's Intralytix—won FDA approval for the first phage product, a food additive that targets listeria in meats.

Why has progress fallen so far short of early hopes? For starters, some scientists have always bristled at the characterization of phages as "utterly harmless." While it's true that phages are ubiquitous (lick your lips and you may pick up a million), to use them therapeutically they must be purified and condensed into a form vastly more concentrated than anything found in nature. Could such a bolus of concentrated virus, however harmless to human cells, trigger a dangerous immune reaction? A few small safety trials suggest not.[51] But there are other potential dangers as well.

"Nearly every kind of toxin produced by disease-causing bacteria stems from a gene delivered by a phage," explains Vincent Fischetti, who has been studying these bacteria-infecting viruses since the 1960s. "That's what phages do. They pick up and deposit genes as they move from one host to the next." The toxins involved in diphtheria, cholera, flesh-eating strep, and hemorrhagic E. coli infections—all are encoded by phages that transformed once-harmless microbes into virulent killers.

Sulakvelidze and many other phage proponents dismiss this danger as easily circumvented. "We search the genomes of our phages for the presence of toxin genes and anything else we don't want to be transferred, and we simply don't use those," he says. Given the millions of phages available in a bucket of polluted water, that task seems simple enough. But it does not stop even the "cleanest" phage from picking up new problematic genes—especially when it's being deployed against superbugs laden with such dangerous DNA. To reduce the risk of this danger, Sulakvelidze adds, one must carefully select phages that have little to no tendency to insert themselves into their hosts' chromosomes.

The problem, Fischetti cautions, is that we're still a long way from fully understanding phages' gene-shuffling behavior. For instance, Fischetti and colleague Thomas Broudy recently showed that toxin-encoding phages behave quite differently in a test tube than they do inside an animal. Place a nontoxic strain of Streptococcus pyogenes into a test tube with a strain that carries a phage-encoded toxin, and nothing happens. But put both strains into the throat of a mouse—or a culture of human throat cells—and the toxin-encoding phage will occasionally jump into the nontoxic strep to transform it into a virulent bug.[52] Though intensely interested in harnessing phages, Fischetti remains circumspect about using them in people. "We simply can't say what will happen when you concentrate these viruses and use them to control a virulent organism," he says.

Beyond safety concerns, phage therapy faces practical hurdles. Because the immune system recognizes phages as viruses, it rapidly clears them from the body. Reportedly, some early Soviet researchers made progress against this problem by repeatedly inoculating animals with a given set of phages till they singled out a few of the viruses that somehow survived in the blood for several days. However, no such immune-evading phages have ever been isolated among those known to target disease-causing bacteria. So even in eastern Europe the successful use of phages has been confined to cures that do not have to be absorbed by the body to work, such as intestinal "flushes" aimed at diarrheal bugs like shigella or salmonella, and phage-impregnated bandages that can be applied to wounds and mucous membranes such as Smeallie's infected sinuses.

One more stumbling block: bacteria rapidly develop resistance against phages, as would be expected from their long history together. So these viruses must be packaged into multiphage "cocktails" that make them both less specific (potentially increasing their collateral damage to the microflora) and more difficult to test for safety. Finally, phages are virtually unpatentable. A company might be able to patent a specific cocktail of phages, but there's nothing to stop another company from mixing up a slightly different one. "Without patents, it's very difficult to get anyone to invest millions into research and development," says Fischetti. "It begins to look foolhardy."

The most biting critique of phage therapy may have come in 2004, when Steven Projan, head of antibacterial research at the pharmaceutical giant Wyeth, noted that a century of phage studies in experimental animals has produced only a single published article showing any benefit. "This silence speaks volumes," Projan wrote in the scientific journal *Nature Biotechnology*. "Indeed the personal, anecdotal testimonials of former patients who 'benefited' from phage therapy are both amusing and sad—we do not hear from those patients whose infections were not cured, for obvious reasons."[53] Parallels to fad treatments such as the debunked cancer cure laetrile become even harder to avoid when reading Internet advertisements aimed at luring patients to Tbilisi for spa cures at the city's growing number of phage therapy centers or at a new satellite clinic "opening soon" in Tijuana, Mexico.[54]

Still, even critics like Projan and cautious proponents like Fischetti can agree that phage research may eventually produce important new tactics for fighting bacterial infections. As a drug developer, Projan sees potential in phage-guided research: phages infect by piercing vulnerable chinks in a bacterial cell's armor, he points out. Why not use them to find new targets for antimicrobial drugs? The Montreal company PhageTech is pursuing this line of attack by analyzing the tiny genomes of staph-infecting phages and studying the proteins they use to hijack a microbe's gene- and protein-copying machinery.[55]

Meanwhile, Fischetti's laboratory at Rockefeller University has isolated a number of enzymes that phages force a bacterial cell to produce once they have finished replicating inside of it. The enzymes perform the bacterial equivalent of hara-kiri, splitting the cell open to release its phage progeny for a new round of infection. Fischetti has found that these "lysins" work equally well when applied to bacteria from the outside. "If they make contact, the bacteria are dead," he says. Even better, most of the lysins that Fischetti has isolated prove to be as highly targeted as the phages that produce them. So far he has isolated lysins effective against anthrax and a half dozen kinds of disease-causing strep bacteria.[56] And he has shown that they work beyond the test tube—for instance, in clearing lab mice of the respiratory bug *Streptococcus pneumoniae*, the strep that causes most human ear infections and bacterial pneumonias. "Just being able to decolonize children and the elderly of pneumococci would save a lot of lives and prevent a lot of otitis media," he says. Best of all, perhaps, Fischetti's pneumococcal lysin doesn't touch "good" strep such as *Strep. salivarius* and *Strep. vestibularis*, which help exclude their more troublesome relatives.

Using phage lysins instead of whole viruses also avoids the dangers of gene swapping and immune reactions. Nor does the immune system clear small molecules such as lysins from the body, so they can linger in tissues much as traditional antibiotics do. In the decade he's been working with lysins, Fischetti says he's rarely seen bacteria develop resistance to them. "We've tried hard to induce it," he says, describing experiments that invite such resistance by exposing bacteria to low concentrations and then searching for partially resistant survivors. "But don't ever underestimate bacteria," he cautions. "It may be more difficult for bacteria to become resistant to lysins than to traditional antibiotics. But it'll happen."

As for when if ever we might see lysin antibiotics, Fischetti is hopeful that he has the investors needed to begin human trials in the next few years. In his favor, he has more than twenty patents on the molecules, including an unusually broad patent on the use of any phage lytic enzyme for preventing infections by "decolonizing" a person's skin, mucous membranes, or intestinal tract of a trouble-prone microbe.[57]

The most promising use for whole phages, Fischetti predicts, will be in eradicating drug-resistant bacteria outside the body, as in hospitals and nursing homes. At Intralytix, researchers are developing such a phage-loaded cleaning product, aimed specifically at listeria. The spray would be ideal, says Sulakvelidze, for use in food-processing plants, particularly for decontaminating known bacterial hotspots such as air-conditioning systems and drains. "Oftentimes even heavy concentrations of chemicals fail to eradicate bacteria from these places," he explains. "The listeria just keep coming back." In these situations, a whole phage might have a decided advantage over its chemical lysins because the virus can persist and multiply as long as it has bacteria to infect.

Intralytix has received a government safety clearance for testing one of its antilisteria cleaning products in poultry-processing plants, and it has already collaborated with the U.S. Agricultural Research Service to test phage sprays and washes on listeria- and salmonella-contaminated produce, with mixed results.[58] Still other research teams have tried using phage to reduce bacterial meat contamination — most successfully by spraying it on the surface of raw beef, poultry, and pork; less so in experiments involving feeding phage to animals right before slaughter.[59]

COCOONS AND FROG SLIME

Around the same time that Western science rediscovered phage therapy, reports of another "natural" bacteria killer were filling scientific journals. Antimicrobial peptides consist of tiny chains of amino acids—like proteins, only smaller. Until 1981 they had been wholly overlooked among the many more complex bacteria-killing chemicals in tears, mucus, and other body fluids. That year the Swedish microbiologist Hans Boman isolated two antimicrobial peptides, or AMPs, from the sleeping pupae of the giant silk moth *Hyalophora cecropia*, naming them "cecropins" in its honor.[60] Boman's cecropins killed a broad variety of bacteria but proved utterly harmless to nonbacterial, or eukaryotic, cells. The exciting discovery helped explain how insects and other invertebrate animals resist infection without the antibodies, T cells, and B cells that empower the more "adaptive" immune systems of higher animals. Four years later, UCLA pathologist Robert Lehrer discovered that people—and likely all forms of multicellular life— make AMPs as well. Specifically, he found them packaged inside the bacteria-gobbling immune cells known as neutrophils. He dubbed these human AMPs "defensins." [61]

The idea of deliberately using AMPs as bacteria-fighting drugs came the following year, in 1986, to a tenderhearted NIH researcher who was in the habit of stitching up his lab frogs instead of discarding them after he'd harvested their eggs. Michael Zasloff had been studying gene expression using the conveniently fat and transparent eggs of the African clawed frog. After anesthetizing a female and surgically removing her eggs, he took a few seconds to crudely close the incision and toss the frog back into the murky tank with her sisters. One day, while pulling several aged, dead frogs from the green water, Zasloff stopped to wonder at the remarkable shape of those that had been under his knife. Though he neither sterilized his scalpels nor cleaned the tank with any regularity, the frogs' incisions had healed beautifully, without the slightest sign of inflammation. Zasloff suspected that the slimy skin of an amphibian must contain a particularly potent version of Boman's cecropins or Lehrer's defensins.

Blenderizing the skin of a few sacrificed frogs, he isolated two antimicrobial peptides that had more broadly powerful antibiotic activity, he claimed, than anything known to science. He called them magainins, from the Hebrew word for "shield," and published his findings to great acclaim in 1987.[62] Like phage therapy, antimicrobial peptides quickly captured the media's fancy. Following a laudatory news article, the *New York Times* editors published an editorial describing Zasloff's work as on par with that of not only Alexander Fleming, penicillin's discoverer, but also Howard Florey and Ernest Chain, the men who had taken a decade to forge penicillin into a working antibiotic. "Dr. Zasloff, aided by the immense power of modern biological techniques, performed all the steps himself, in a single year," the editors wrote, noting that the discovery could not have come at a better time to save humankind from the escalating crisis in antibiotic resistance. "Even if only part of their laboratory promise is fulfilled," the editorial concluded, "Dr. Zasloff will have produced a fine successor to penicillin."[63]

Further study revealed how antimicrobial peptides wreak their selective destruction: bearing a mildly positive electric charge, they cling to the negatively charged outer surface of a bacterial membrane but not to the relatively uncharged membranes of animal cells. Once stuck to the microbe's surface, the peptides change shape in a way that pierces the cell envelope. Peppered with such holes, the bacterium leaks to death, though in reverse fashion — with an inrush of water from its surroundings.

Zasloff felt he had discovered a true microbial Achilles' heel. "Despite their ancient lineage, antimicrobial peptides have remained effective defensive weapons, confounding the general belief that bacteria, fungi, and viruses can and will develop resistance to any conceivable substance," he announced.[64] For bacteria to become resistant to these peptides, it seemed, they would have to fundamentally change the physical structure of their membranes to reverse the electrical charge, something Zasloff and others argued was next to impossible. Such hubris seemed to tempt the gods. But Zasloff's studies appeared to back up the claim.[65]

Zasloff and his investors wasted little time forming a private company to fund patient trials. As a hospital pediatrician as well as a researcher, Zasloff was particularly interested in developing AMP treatments for young cystic fibrosis patients. In 1997, while at the University of Pennsylvania, he had helped demonstrate that the disease appeared to result, at least in part, from defective defensins in the lungs.[66] Meanwhile, Zasloff's early reports had set off a land rush of discoveries and patents for other antimicrobial peptides, each one a potential billion-dollar molecule in a world eager for new antibiotics.

By 1998 Zasloff had tested his magainins in more than a thousand volunteers as a treatment for impetigo, a bacterial skin infection, and for diabetic skin ulcers; in both cases the AMPs proved modestly effective in relieving or preventing infection.[67] That was enough to gain the interest of the pharmaceutical giant SmithKline Beecham, which wanted to market Zasloff's AMP antibiotic cream as the new drug Locilex.

The following spring, the sprint to market came to an abrupt halt when the FDA advisory board considering final approval announced that, while it was satisfied that Locilex was safe, it wanted more studies on its effectiveness.[68]

The decision sparked outrage among hopeful diabetes patients and an abrupt loss of interest on the part of SmithKline Beecham. But related AMP research continued apace. In fact, so great was his confidence in AMP's safety that Zasloff proposed developing another of his company's products—the shark AMP squalamine—as an appetite suppressant. Tests in lab mice had unexpectedly shown that the peptides prompted animals to stop eating.[69] By 2001 the number of newly discovered antimicrobial peptides was nearing five hundred, and thousands of related papers had appeared in the scientific press. That same year, Michael Shnayerson and Mark Plotkin published *The Killers Within*, a doomsday account of the deadly rise of drug-resistant bacteria. In the book's concluding chapters these respected science journalists gave antimicrobial peptides equal weight with phage therapy as the two great hopes for saving modern medicine.

It took a pair of evolutionary biologists to throw cold water on the AMP fire. In June 2003, Graham Bell, of Canada's McGill University, and Pierre-Henri Gouyon, of the University of Paris, published an opinion piece titled "Arming the Enemy" in the journal *Microbiology*.[70] However safe AMP drugs may appear in the short term, they warned, the consequences could be disastrous should their use breed resistance. "The evolution of resistance to any antibiotic makes it less useful in treating disease, of course," they wrote. "It also deprives any organism that produces it of part of its antibacterial armory. This would not normally be a matter for concern; but in the case of antimicrobial peptides, *we ourselves are the producers*." In theory, resistance to AMPs could create bacteria that defied the very chemicals that the human body uses as its first line of defense. The potential side effects: minor cuts that no longer healed; eye infections and lung disease caused by germs in ordinary air. Even the most innocuous of the body's resident bacteria could become an invasive threat.

Bell and Gouyon's warning produced a stunned, then angry response from the many prospectors in the new pharmaceutical field of antimicrobial peptides. The first to reply in the scientific press, Zasloff dismissed the possibility of resistance as "improbable" and called Bell and Gouyon's logic "fundamentally wrong."[71] Bacteria have had millions of years to develop resistance to these chemicals, he argued, and it hasn't happened. In a personal challenge, Zasloff dared Bell to use pexiganan, the most widely tested of his magainin antibiotics, on any microbe at any concentration for any length of time. "I'll bet this peptide will not elicit resistance," he told reporters. Bell accepted the challenge, and the two men agreed to publish the results under both their names.

Out of twenty-four bacterial cultures grown in Bell's lab under the influence of pexiganan, twenty-two became resistant to the drug.[72] How did an evolutionary biologist succeed in producing resistance where a microbiologist such as Zasloff had failed? Bell and one of his undergraduate students, Gabriel Perron, used a time-honored method for selecting and fostering resistant organisms. They began by growing their bacterial strains (a dozen kinds of E. coli and a dozen pseudomonas) in a broth inoculated with an extremely low dose of pexiganan, virtually guaranteeing a few survivors. Each morning, Perron had the task of isolating the survivors and transferring them to a fresh test tube. He doubled the pexiganan dose every few days. In this way the rare surviving mutant had time to accumulate additional mutations that might increase resistance. The process, though laborious to perform in the laboratory, was not unlike what a bacterium might experience as it spreads and multiplies in bodies and other environments contaminated with any type of antibiotic.

An evolutionary biologist, not a geneticist, Bell left it to others to decipher exactly what mutations had enabled the AMP-bred microbes to survive. His point had been made: though bacteria may not develop resistance to antimicrobial peptides in nature, deploying these chemicals in a concentrated, prolonged manner—as one does when treating an infection—changes evolutionary pressures in a fundamental way. As for Zasloff, he has embraced the results with remarkable candor for one who had dedicated nearly twenty years of his career to what he firmly believed was a resistance-proof drug.

"If something can happen in a test tube, it is very likely that it can happen in the real world," he announced on the eve of the report's release.[73] At the same time, Bell joined Zasloff in affirming that their results should not discourage research into antimicrobial peptides, but rather should serve as a wake-up call for researchers to proceed more cautiously with the dozens of AMP drugs nearing medical and veterinary use. In addition to more rigorous testing for resistance, they urged scientists to further explore the danger of cross-resistance: that is, to exhaustively test whether AMPs from animals such as insects, fish, or frogs can produce cross-resistance to the AMPs produced by humans, their pets, and their livestock. Such cross-resistance has long been a problem with conventional antibiotics produced by organisms as distantly related as fungi and bacteria (two entirely different kingdoms of life).

A safer, more promising approach may lie in treatments aimed at boosting the body's own levels of antimicrobial peptides. Both Zasloff and Bell have endorsed this approach, which would be less likely to promote resistance because the body deploys AMPs in combination with many other defenses such as antibodies and antibacterial enzymes. In the first animal test of such a treatment, Swedish and Bangladeshi researchers recently treated shigella-infected rabbits with sodium butyrate, a fatty acid salt normally found in trace amounts in the colon. In and of itself, butyrate has no antibiotic properties. However, it gooses the intestinal tract's production of cathelicidin, a potent antimicrobial peptide that prevents bacteria from attaching to the delicate cells of the intestinal lining. In the rabbit experiments, those given butyrate began recovering from their shigella dysentery within the day, while the condition of a comparison group given a placebo was clearly worsening.[74]

In a related study with human immune cells, UCLA's Robert Modlin found that vitamin D boosted their production of defensins along with their ability to clear tuberculosis bacteria from a cell culture.[75] The finding may explain the benefit of the early twentieth-century "sunshine cures" practiced at TB sanitariums, as sunlight stimulates the skin's natural production of vitamin D. Modlin's team also showed that blood serum from people of African ancestry, a group known to be particularly vulnerable to tuberculosis, has lower levels of both vitamin D and antimicrobial peptides—most likely due to the sunlight-shielding power of darker skin pigment. Modlin has now advanced his test-tube findings to clinical trials to see if inexpensive vitamin supplements can bolster TB resistance in dark-skinned peoples. The finding is no trivial matter in light of the recent appearance of virtually untreatable strains of XDR TB, or extreme-drug-resistant tuberculosis.[76]

Meanwhile, interest in developing antimicrobial peptides into antibiotic drugs continues despite their newfound controversy. If cross-resistance to human AMPs turns out to be a false concern, then AMPs may safely and significantly extend the golden age of antibiotics. But Bell and Perron's hundred-day experiment reminds us that anything that kills bacteria inevitably promotes the rise and spread of resistance. Given this hard reality, it seems foolhardy to hope that even the most prudent use of existing antibiotics and the rapid development of new ones will do more than delay the next crisis in drug resistance.

As naïve as it may sound in a day when killer superbugs dominate our headlines, a growing scientific consensus is forming that it's time to move beyond our escalating war on microbes and look for ways to foster a truce in what will always be a bacterial world. More than an idealistic philosophy, research in this vein has begun to produce results.

PART VI

BEYOND LETHAL FORCE—
DEFANG, DEFLECT, AND DEPLOY

Our indigenous biota is, in fact, part of the environment in which we live. We need to accept its existence. Acceptance, however, need not be passive or re-signed: the biota is no less subject than the rest of our environment to manipulation for human benefit.

—microbiologist THEODOR ROSEBURY, 1962

DRUGS THAT DISARM

It was the kind of experiment no scientist enjoys: injecting bacteria under the skin of a mouse to produce a nasty abscess. But the goal was a huge one: to stop the kind of tissue destruction that has taken countless human lives and limbs in the ages since man and staph began coexisting in a state of alternating truce and warfare—and to do so not by trying to kill these slippery bacteria, but by defanging them.

In the fall of 2004, Richard Novick was close to pulling off the trick. A grizzled and gravelly-voiced veteran of the microbe wars, Novick had spent the previous twenty years at the head of a New York University lab devoted to deciphering how *Staphylococcus aureus* performs its infamous Dr. Jekyll–to–Mr. Hyde transformation: one day it resides harmlessly on a person's skin or mucous membranes; the next day it unexpectedly breaches all barriers and causes life-threatening bloodstream infections and organ damage. Novick had shown, for example, how staph switches on a suite of more than a hundred genes involved in churning out cell-splitting, bloodstream-clogging, organ-destroying toxins.

In theory, a drug that blocked staph's transformation from harmless to virulent would be far less likely than antibiotics to breed resistance, for it would leave the bug's eradication to the body's multifaceted immune response. It should also prove utterly harmless to the well-behaved members of the body's protective microflora.[1]

The guiding principle behind this new approach to fighting bacterial disease is an updated version of Koch's famous postulates, which a hundred years ago focused medical science on the task of determining which microbial species caused what diseases. Today's molecular version shifts the search to a subtler aspect: What is it about a particular strain or species that enables it to cause harm?

The impressive way that staph coordinates its assault showed Novick how to proceed in developing his virulence-blocking drug. For *Staph. aureus* has the timing of a savvy field general. It does not attack too soon—which would draw an immune response before its cloned army was large enough to fight. Nor does it wait too long—and risk being detected and routed before it has drawn its weapons. So how does staph "know" when it has amassed enough strength to fight?

Novick understood that staph, like many kinds of bacteria, can take a crude census of its numbers with small molecules known as auto-inducing peptides, or AIPs. These peptides diffuse away from a small group of bacteria, but once a colony reaches a certain size and density, they start accumulating. In essence, this accumulation of molecules tells the colony when it has a "quorum," a number sufficient for group activity. To block such a signal, Novick reasoned, would effectively keep staph's weapons holstered.

Best of all, Novick discovered that staph itself provides the means to accomplish this holstering. All that was required was a little human trickery. For Novick and his students had discovered that each different strain of *Staph. aureus* produces one of four different autoinducing peptides, prosaically dubbed AIP-1 through AIP-4.[2] Flooding staph from one group with the signaling molecules of another might effectively jam its virulence switch in the off position.

And so, in the fall of 2004, Novick and his students injected six hairless lab mice with an AIP-1 strain of *Staph. aureus*. Three of the mice also received a shot of AIP-2 peptides. Over the next week, the mice injected with staph alone developed large boils at the site of their injections, which then burst to leave gaping abscesses. By contrast, those that also received the mismatched peptide suffered little more than small blisters that drained and healed over.[3] "What we've done is simply tip the balance in favor of the immune system," Novick says. "It's not that staph can't cause disease without switching on its toxin production. But bacteria caught without toxins may not be able to block or kill the neutrophils that arrive to clean them up."

Novick has now moved beyond testing his virulence-blocking drugs in mice to working with human cell cultures and deadly strains of staph isolated from the lungs of hospital patients. Severe bacterial pneumonia kills around one in four patients who require ventilator-assisted breathing following a stroke or heart attack.[4] Could an infusion of the right virulence-blocking peptide prevent these deaths? "In principle it's a great idea," says Novick. But many questions remain to be answered, beginning with how quickly these chemicals need to be administered to be effective.[5] Novick and his team are now working to resolve some of these challenges.

Meanwhile, two hundred miles to the north, at Harvard Medical School, Michael Gilmore pursues a related set of microbe-taming tricks. His challenge: vancomycin-resistant *Enterococcus faecalis* (VRE). Many strains of this hospital superbug produce toxins called cytolysins that destroy other types of cells, including human blood cells and even other kinds of intestinal bacteria. This destructive lifestyle, combined with an impressive degree of antibiotic resistance, makes VRE truly formidable. "Compared with the garden-variety enterococci we all carry in our intestines, the hospital strains that express cytolysins are around a hundred times more lethal," says Gilmore.

Like staph, VRE does not deploy its weapons randomly. Gilmore and his collaborators have discovered that VRE's cytolysins double as a kind of chemical radar that tells the microbe when it comes within range of its quarry. VRE continually secretes its cytolysins, ordinarily at low levels and always in two parts: a long molecule and a short one, both of which must bind to a human cell to destroy it. "So here's the elegant trick," Gilmore says of his discovery. "The two subunits bind to each other when they do not find target cells in their vicinity. And when there is a target, the long subunit binds to it much more quickly than does the short one." So a sudden spike in the amount of "short" cytolysin effectively tells VRE that its victim has just come within range. In response, it immediately increases its toxin production more than a hundredfold.[6]

While this clever turn enables VRE to conserve its ammunition until needed, it also provides Gilmore with the ideal target for a virulence-blocking drug. "The short unit," he says, "is essential both for killing human cells and for signaling *E. faecalis* to turn on cytolysin production full blast." He and his colleagues have designed chemical inhibitors that bind strongly and quickly to this crucial peptide to render VRE blind to the human cells that it would otherwise kill. Like Novick's staph-jamming AIPs, Gilmore's drug doesn't kill its target. It just hobbles one of its most dangerous weapons. That's particularly important given the bounty of harmless and potentially protective strains and species of enterococcus residing in a patient's gut. "Today we're just blasting away with antibiotics," he says, "and that's exactly what promoted the rise of these resistant and virulent strains."[7]

Strep bacteria present a third important target for a nonlethal approach to fighting infection. At the University of California, San Diego, pediatrician Victor Nizet has found several intriguing targets for blocking the virulence of both *Streptococcus agalactiae*, the group B strep that causes deadly infections among newborns, and *Streptococcus pyogenes*, the bacterium that causes strep throat and so-called flesh-eating skin infections. With group B strep, Nizet has homed in on a single gene whose activity produces what he calls strep's "sword and shield." The shield consists of a pigment that protects the bacterium from destruction when it ends up inside a microbe-gobbling immune cell. The sword is a toxin that destroys these same soldier cells by riddling them with holes.[8]

With *Strep. pyogenes*, Nizet and his students have deciphered the biochemical steps that the bacterium uses to produce streptolysin, its signature toxin. They have also discovered that it produces a novel enzyme that allows it to slice its way out of a netlike trap that some immune cells cast over their prey.[9] These and other virulence factors are now the targets of Nizet's research efforts and, he hopes, the prototypes for a new generation of antibiotics.

While most research into "defanging" bacteria remains on the laboratory bench, the Massachusetts biotech company Genzyme may be close to bringing the first such drug to market. Genzyme's tolevamer is a gummy molecule that binds to the intestine-destroying toxins produced by one of today's most dangerous superbugs, the hypervirulent new strain of *Clostridium difficile*. Once doctors can block those toxins, it becomes safe for them to withdraw the antibiotics they normally use to force *C. difficile* to retreat into its drug-resistant spores. And *that* allows the patient's normal intestinal microflora to rebound to levels that can crowd out the problematic bug on a permanent basis.

This sort of toxin-binding drug has a long tradition in the treatment of gastrointestinal infections, going back to activated charcoal and bismuth (the active ingredient in Pepto-Bismol). In essence, tolevamer is a more potent and targeted version of such over-the-counter remedies.

It also offers a huge advantage over the now-standard approach of fighting *C. difficile* infection with microflora-razing antibiotics—which, after all, are what triggers most *C. difficile* infections.

As of early 2007, more than sixty medical centers were using low doses of tolevamer on an experimental basis. In early trials the drug proved comparable to the antibiotic vancomycin in resolving *C. difficile* diarrhea and colitis and slightly better at effecting a lasting cure— that is, in preventing rebound infections.[10] With these early trials showing the drug safe, the participating medical centers are now using tolevamer in higher doses with hopes of better results.[11]

VACCINES—FOREWARNED IS FOREARMED

Just as toxin-binding drugs can disarm bacteria, the right vaccine can prompt the immune system to produce antibodies that do the same. Some of medicine's oldest and most effective vaccines target virulence in this way. The tetanus vaccine, for example, induces antibodies that neutralize the powerful muscle-contracting toxin tetanospasmin. The diphtheria vaccine protects against a bacterial poison so lethal that eight-millionths of a gram can kill a full-grown man. When widely used, virulence-targeting vaccines can have the added benefit of promoting the rise of "gentler" strains of bacteria that do not waste their energy producing weapons to which their hosts have become immune. That has been the case, for example, wherever countries have adopted near-universal immunization with the diphtheria vaccine.[12]

Beyond being a way to defang bacteria, vaccines may be our best hope for reducing both our lifetime burden of disease-related inflammation and our incessant use of antibiotics. Vaccines turn out to be far less likely than antibiotics to promote resistance, as they do not target a particular structure (such as a cell wall) or molecule (such as ribosomal RNA) but instead prime the immune system to rapidly clear a given bug with a multifaceted attack.

Today's list of vaccine-preventable diseases includes those caused by a dozen-odd viruses and a half dozen bacteria. The latter include tetanus, diphtheria, whooping cough, and bacterial meningitis. Among the newest, the *Streptococcus pneumoniae* vaccine introduced in 2000 slashed rates of pneumococcal pneumonia and meningitis in the United States from more than 60,000 cases a year in the 1990s to 37,000 a year (and dropping) in 2002.[13] As an added bonus, the vaccine produced an abrupt reduction in drug resistance in the pneumococcal infections it failed to prevent. This is because five of the seven *Strep. pneumoniae* strains targeted by the vaccine had been responsible for 80 percent of drug-resistant infections.[14]

Clearly, we're far from having a vaccine for every bacterial disease that plagues us. "We long ago developed all the easy ones," says Henry Shinefield, whose work battling staph in hospital nurseries of the 1950s led to a career in vaccine research. Some bacteria evade easy capture by vaccines, he explains, because they come in myriad strains that present different "faces" to the immune system. Other bacteria somehow avoid provoking lasting immunity even after an active infection. This is especially true of bacteria that hide their surface proteins from the immune system inside a capsule made of polysaccharides. For proteins provoke by far the strongest and longest-lasting types of immune responses.

All these factors and more have bedeviled those pursuing immunology's holy grail of an effective staph vaccine. After decades of concerted effort, Shinefield and his associates at Kaiser Permanente Vaccine Study Center, in Oakland, California, have come closest. In the 1990s they developed StaphVAX, a vaccine that combined pieces of staph's polysaccharide surface capsule with proteins designed to evoke a strong immune response. In 2002 they announced the results of a StaphVAX trial involving more than eighteen hundred dialysis patients, a group at high risk of deadly staph blood infections.[15] The vaccine halved the patients' risk of developing staph infections, but the partial immunity lasted only nine months, after which the protection rapidly waned. In a high-risk group such as dialysis patients, this level of protection might be well worth a yearly or twice-yearly jab. But it's far from delivering a practical staph vaccine for the rest of us.

New hope comes out of twenty-first-century advances in immunology and gene technology. In particular, the gene sequencing of potentially harmful bacteria opens up a new way to search for the ideal molecules to include in a vaccine to maximize protection. At the University of Chicago, for example, microbiologist Olaf Schneewind and graduate student Yukiko Stranger-Jones are using a technique dubbed reverse vaccinology in their quest to deliver a vaccine against North America's most dangerous and prevalent strains of methicillin-resistant *Staph. aureus* (MRSA).

In traditional vaccine design, researchers biochemically dissect a bug and then combine various bits and pieces to try to find those that provoke the most protective immune response in laboratory animals. By contrast, reverse vaccinology launches the search for targets with a computer program. Stranger-Jones used it to screen the genomes of eight different strains of MRSA to find their common motifs. From these she identified nineteen potential vaccine targets—genes that coded for common surface proteins. Isolating the proteins, she then tested them individually, by injecting each protein into mice to see how well exposure protected the animals from later infection with live staph bacteria. The top four candidates included two proteins that help staph capture the iron it needs (from red blood cells) and two that probably help the bug adhere to human tissues. Individually these proteins produced only weak protection in lab mice. However, when Stranger-Jones injected mice with all four, she rendered the animals completely immune to two strains of virulent MRSA and produced partial immunity against three other strains.[16] "This is just the start," she says. Early 2007 found Stranger-Jones delving back into her staph genomes to find other common targets to boost her vaccine's power.[17]

Taking a different tack, researchers at the California pharmaceutical company Cerus are packing vaccines with live bacteria that can infect cells but not replicate inside them. The tactic may prove to be a breakthrough for fighting diseases such as TB, typhus, chlamydia, brucellosis, and listeriosis, which are caused by bacteria that hide inside human cells. To fight these intracellular infections, the immune system has to generate antibodies not to the bacteria per se but to the cells they infect. The compromised cell aids in this effort by marking itself for destruction—using bits of bacterial proteins posted on its surface. For this reason, vaccines containing weakened but still infectious bacteria work much better against intracellular bugs than do vaccines that contain either dead bacteria or their component parts, which remain outside our cells. Unfortunately, live vaccines come with risks, especially for the immune compromised, whether they receive the vaccine themselves or simply come in contact with those who do. Cerus micro-biologist Tom Dubensky came up with the solution of effectively neu-tering an intracellular pathogen—in this case, the food-poisoning bug *Listeria monocytogenes*. He did so by knocking out several of the bac-terium's genes that are essential for DNA repair. He then irradiated the compromised bacteria with ultraviolet light. The result: listeria cells that could do virtually everything a normal listeria bug could do except reproduce.[18]

These and other new approaches are reinvigorating hopes that vac-cines may someday deliver the victory against infectious disease that so many thought was at hand with the discovery of antibiotics. But simply bringing a new vaccine to market is not enough, warns Tufts's Stuart Levy. To be sure, vaccines seldom induce the kind of resistance pro-duced by antibiotics, prompting mutations or gene acquisition that render the vaccine ineffective. However, vaccines that target antigens found on some but not all strains of a bacterium can promote the rise of strains not included in the vaccine. That's been the case, for example, with the pneumococcal vaccine introduced in 2000, which targeted the seven most common out of hundreds of strains of *Streptococcus pneu-moniae*.[19] Staying ahead in this game, Levy says, requires a commit-ment by national and international health agencies to continually monitor what bacterial strains are active in their countries and commu-nities, and a matching commitment by vaccine makers to periodically rejigger their vaccines to reflect any changes.

As vaccine-preventable diseases go from commonplace to rare, health officials will also need to adjust their recommendations for booster shots—to make up for the natural booster effect once provided by exposure to infected family and friends. The recent comeback of whooping cough, for example, has resulted in part because of the wan-ing immunity of adults whose only encounter with the bug was their own childhood immunization. It's now clear that lasting immunity to pertussis requires a booster shot in late adolescence and perhaps an-other jab in middle to late adulthood.[20]

Finally, warns Levy, microbiologists must remain vigilant in moni-toring the effects of any vaccine that targets a member of the body's normal microflora. "Whenever we knock one organism out of its niche, we have to remember that something else is going to take its place," he explains. As a prime example, he points to the vaccine against *Haemophilus influenzae* type b (Hib), a common denizen of the nose and throat. Before the introduction of effective vaccines (for toddlers in 1987 and for infants in 1990), Hib was the leading cause of bacterial meningitis, causing some twenty thousand infections and nearly a thousand deaths a year in the United States.[21] The Hib vaccines slashed those infection and death rates by more than 80 percent.[22]

While celebrating the news, a few chary microbiologists such as Levy waited to see what would take Hib's place. The less savory possibilities included *Strep. pneumoniae* or *Staph. aureus*. Hib's replacement emerged as a set of nontype-b haemophilus bacteria that occasionally cause sinusitis, primarily in adults. "So we ended up trading a life-threatening disease for a life-bothering one," says Levy. "Not a bad trade, but we may not always be so lucky."

DOMESTICATE AND DEPLOY

Rather than leave the result to chance, Levy and many others are interested in proactively replacing the body's trouble-prone bacteria with strains and species of our choosing, even of our own making.

Call it probiotics, or competitive exclusion, or bacterial-replacement therapy—the concept of deliberately improving the quality of our microbial tenants dates back to Elie Metchnikoff, the same nineteenth-century microbiologist who mistakenly viewed our intestinal microflora as unmitigated parasites. Metchnikoff may have been wildly misguided in wanting to surgically remove our colons (just to rid us of their noxious bacteria), but he appears to have been on the right track in advocating the daily consumption of fermented drinks and cheeses containing lactic-acid bacteria. Metchnikoff believed that these "good" bacteria waged battle with our native microflora. We now know that they work *with* our intestinal bacteria to repel would-be invaders such as rotaviruses, listeria, salmonella, and other gastrointestinal bugs.[23]

The most thoroughly studied of modern-day probiotics is a supplement comprised of an unusually tenacious strain of the bacterium *Lactobacillus rhamnosus*, dubbed *Lactobacillus GG* in self-designated honor of its developers, Tufts University's Sherwood Gorbach and Barry Goldin. Starting in the early 1960s Gorbach had worked with the dairy industry to identify the most healthful kinds of milk-digesting bacteria. In particular, he was interested in finding strains and species that could take up lasting residence in the human colon. Over twenty years of research, however, he failed to find anything remotely tenacious among the bacteria in American, European, or Asian dairy cultures. Though most live food bacteria perish in the acid bath of the stomach, many of these milk-digesting bacteria reached the colon alive. Nonetheless, they disappeared within a day or two.

Then, in 1983, Goldin joined Gorbach's lab team, and the two men decided to abandon research on dairy bacteria and, instead, search for beneficial lactobacilli in the human digestive tract. Their new tack required the cooperation of fellow scientists, family, and friends, from whom Goldin cajoled donations of stool.[24]

As they began screening this harvest, Gorbach and Goldin drew up a wish list of desirable traits. The ideal candidate for a probiotic supplement, they decided, would consistently survive a bath of stomach acid and intestinal bile, adhere strongly to a column of laboratory-cultured intestinal cells, and be able to elbow its way around a petri dish filled with normal intestinal bacteria such as E. coli, as well as more dangerous gastrointestinal bugs such as salmonella.

In the spring of 1985 they isolated a bacterium that scored reasonably well on all three tests: Lactobacillus GG, now sold as the nutritional supplement Culturelle. In the two decades since this strain's discovery, Gorbach, Goldin, and many others have published more than a hundred scientific papers demonstrating its benefits, most clearly its ability to help prevent and relieve gastroenteritis, the intestinal irritation and inflammation that can result from either infection with a gastrointestinal bug or the microflora-disruption of antibiotics.[25]

Beyond the intestinal tract (if only by an inch), probiotics are also proving their ability to protect against common vaginal and urinary tract infections and reduce the risk of sexually transmitted disease. In the early 1970s Canadian urologist Andrew Bruce showed that women who get recurrent vaginal and urinary tract infections tend to have stray E. coli in their vaginas, while the vaginal microflora of women who seldom if ever get these infections consist of a select group of lactobacilli.

These lactobacilli appeared to aggressively discourage the incursion of interlopers from the nearby intestinal tract.[26] An abundance of follow-up studies confirmed that a healthy vagina was one dominated by lactobacilli.[27]

In the 1980s, Gregor Reid, of the University of Western Ontario, furthered Bruce's work with a search for a vaginal equivalent to *Lactobacillus GG*. Like Gorbach and Goldin, Reid found that dairy lactobacilli such as *L. acidophilus* didn't have the right stuff to take up residence where he wanted them.[28] Nor did *Lactobacillus GG*. He began collecting bacterial swabs from the vaginas of women who hadn't experienced a vaginal or urinary tract infection in several years. From hundreds of possible candidates, he identified two strains that strongly repelled intestinal bacteria in the lab. *Lactobacillus rhamnosus* G-1 and *L. fermentum* RC-14 beat back would-be competitors with an abundance of hydrogen peroxide and a variety of biosurfacants (slippery molecules that make it hard for other bacteria to get a grip). These two lactobacilli had the added advantage of not being fazed by spermicides, which have a nasty tendency to raze vaginal bacteria and so predispose to infections.[29]

In testing his probiotic, Reid followed the health of more than a hundred women who either swallowed a capsule of the supplement or inserted it as a vaginal suppository. In either case, the bacteria in the probiotic reach a woman's vaginal tract, become the predominant residents there, and so restore the kind of lactobacilli microflora associated with resistance to infections.[30] And in a study with forty women suffering from bacterial vaginosis, the probiotic proved itself a more effective cure than the standard treatment of metronidazole antibiotic gel. Vaginal flora returned to normal in eighteen of the twenty women taking the probiotic, a cure rate of 90 percent. By comparison, metronidazole cured just over half—eleven out of twenty women.[31] Reid published the results of the small clinical trial in 2006, the same year that his probiotic debuted in U.S. and Canadian health food stores as FemDophilus.[32]

Unfortunately, for every scientifically tested probiotic standing on the shelves of vitamin shops and drugstores, there are dozens of outwardly similar products filled with dubiously useful and sometimes mislabeled organisms.[33] Most are probably harmless. But some contain antibiotic-resistant bacteria—a huge concern to microbiologists who understand how readily resistance genes can spread from a probiotic to a person's intestinal microflora and from there to disease-causing organisms. Some probiotic companies go so far as to tout the antibiotic resistance of their products.

PRESCRIPTION PROBIOTICS

As a young doctor in the early 1980s, Kristian Roos often puzzled over the stubbornly recurrent strep throat infections that plagued many patients at Sweden's Gothenburg University Hospital. Some patients turned up in his clinic several times a year, their flame-red tonsils and throats speckled with white patches. Roos knew that around a quarter of us continue to harbor *Streptococcus pyogenes* in our throats even after antibiotic treatment knocks back its numbers enough to stop active infection. This persistence partially explained why some people remain predisposed to reinfection. But it didn't explain why some, but not others, become persistent carriers.

Could it be that *Strep. pyogenes* faced less competition in the throats of tonsillitis-prone patients than it did in those who remained healthy? In 1985 Roos took the helm of the ear, nose, and throat clinic at Gothenburg's Lundby Hospital, where he had a better chance of answering this question. At Lundby, Roos would see more of the routine infections that interested him. Just as important, the clinic's regular schedule of wellness checkups gave him the chance to take throat swabs from the healthy as well as the sick.

Sure enough, Roos found that while most people have an abundance of harmless alpha-streptococci in their throats, those who carried *Strep. pyogenes* tended to have little to none. Perhaps the "good" strep were literally keeping their troublesome cousin in check. The experience of one family backed up Roos's hunch. After the youngest child developed a chronic *Strep. pyogenes* skin infection, the mother began suffering recurrent strep throat. Strain typing showed that the exact same substrain infected them both, but the boy never developed tonsillitis. The difference, Roos found, was that the boy carried massive amounts of alpha-streptococci in his throat, while his mother lacked them entirely.[34]

Roos and several colleagues began studying the many kinds of streptococci bacteria that grew in the throats of the healthy. By 1995 they had packaged several of these bugs into a throat spray. In a preliminary study, patients being treated for strep throat received either the probiotic mix or a dummy saltwater spray to use daily for a week after finishing their antibiotics. Over the next nine weeks, only one of the 51 probiotic users suffered a second bout of strep tonsillitis, compared with 14 of the 61 patients who used the placebo—a tenfold difference in lasting cure.[35] In a larger study of 342 patients followed for over ten weeks, the difference proved less dramatic but significant: tonsillitis returned in less than 20 percent of those who spritzed their throats with the live alpha-strep bacteria, compared with 30 percent of those who got the saline spray. At the end of the study, those using the probiotic were also half as likely to silently carry Strep. pyogenes in their throats as were those who received the placebo.[36]

While he was hoping and waiting for a pharmaceutical company to develop his probiotic throat spray, Roos began wondering whether a lack of protective bacteria might likewise contribute to the recurrent ear infections that plague so many toddlers. Stray throat bacteria cause these infections when they become trapped inside a child's middle ear chamber. Knowing that some throat bacteria cause more ear trouble than do others, Roos began looking at what lived in the upper throats of healthy children. In the process, his research team amassed a collection of some six hundred kinds of alpha-streptococci, which they tested and ranked according to their ability to inhibit the four kinds of bacteria that most frequently turn up inside children's infected ears: Streptococcus pneumoniae, Haemophilus influenzae, and to a lesser extent, Strep. pyogenes and Moraxella catarrhalis.

In 1996 Roos packaged the five most ear-protective bacteria into a nasal spray, which he gave to the parents of toddlers with a history of chronic infection. In a study of 108 children, half received the probiotic spray once daily for ten days; half received squirts of salt water. At the end of three months, nearly half of those who got the probiotic remained free of ear infections. The same could be said of less than a quarter of those who got the dummy spray.[37]

Though few outside Scandinavia ever heard of Roos's strep throat studies, the rise of stubbornly drug-resistant ear infections ensured that his nasal "bug spray" made international headlines when the *British Medical Journal* published the results in January 2001.[38] The year before, the American Academy of Pediatricians had, for the most part, thrown in the towel on using antibiotics to treat ear infections. At the time, studies were confirming that the drugs helped little and clearly predisposed babies and toddlers to antibiotic-resistant respiratory and intestinal infections.[39]

Nonetheless, Roos has found little corporate interest in developing his ear-protective probiotic. The problem may lie in profitability, he admits. Unlike Culturelle or FemDophilus, which are marketed as "nutritional supplements," probiotic sprays intended for medical use would have to pass through an expensive gauntlet of clinical trials to prove their safety and efficacy. For this kind of multimillion-dollar investment, investors would want to know they held exclusive rights to the treatment. "But even though we can patent our particular mixture of organisms," Roos says, "it would be so easy for someone else to come along and put together something slightly different from the hundreds of protective strains found in people's throats."

On yet another front, the last few years have produced an about-face in the use of probiotics for the prevention and treatment of *Clostridium difficile* colitis. As recently as 2001, U.S. gastroenterologists continued to dismiss such treatments as quackery, despite their growing use in Europe and Australia.[40] Then came the deadly rise of hypervirulent *C. difficile*. A 2006 reanalysis of more than thirty placebo-controlled clinical trials, many of them European, confirmed that at least two commercially available probiotics—*Lactobacillus* GG and the "baker's yeast" *Saccharomyces boulardii*—help prevent *C. difficile* infection when taken during and immediately after a course of antibiotics. Saccharomyces also proved effective in treating patients who were already suffering from *C. difficile* disease, reducing by almost half the rate of recurrence following standard treatment with the antibiotic metronidazole or vancomycin.

A more effective probiotic cure for *C. difficile* colitis may be in the works following successful animal tests with several nontoxic strains of the same bug. The benign *C. difficiles* come from the five-thousand-plus strain collection of the Northwestern University microbiologist Dale Gerding, who in 2005 identified the hypervirulent new strain that was killing hospital patients across North America and Great Britain. Gerding selected three nontoxic strains from his collection based on their frequent appearance in the stool of hospital patients who did *not* develop diarrhea or colitis while others around them did. In 2002 Gerding showed that inoculating hamsters with any one of the three benign strains protected over 90 percent of the animals from subsequent inoculation with toxic strains.[41] As *C. difficile* does not take up permanent residence in the gut, it is unlikely that the benign strains are competing with the toxic ones. Rather, Gerding believes that they may act as a live vaccine that raises protective antibodies against subsequent infections. Since 2006 he has been working with the biopharmaceutical company ViroPharma (in Exton, Pennsylvania) to test his protective strains in patients.[42]

FIGHTING FIRE WITH FIRE

When Heinz Eichenwald and Henry Shinefield opened up the controversial field of "competitive exclusion" in the 1950s, they had the audacity to deliberately inoculate babies with *Staph. aureus*, albeit a relatively benign strain of this infamous bug. Today there's no doubt that their 502A staph strain saved the lives of scores, possibly hundreds, of newborns—if only due to the virulence and drug resistance of the 80/81 staph strain that was then running rampant in hospital nurseries.[43]

As recently as the early 1990s, at least one doctor was still using 502A to combat persistent and virulent staph infections. Immunologist Russell Steele's rediscovery of 502A followed his 1978 arrival at the University of Arkansas Medical School, in Little Rock. "An immunologist was kind of a rare bird in those days," he recalls. "You were supposed to know why people got sick in the first place. So the local dermatologists began sending me their most difficult cases." These cases included entire families who had been plagued with the yellow boils and angry red abscesses of staph skin infections for months, sometimes years. A search through the medical literature led Steele to Shinefield, who encouraged him to request a vial of 502A from the national repository of the American Culture Collection, in Rockville, Maryland.

Unlike the "virgin territory" of Shinefield's newborns, Steele's patients were already stubbornly colonized by staph. So he began his treatments with a barrage of oral antibiotics, prescription ointments, and twice-daily showers with the potent disinfectant hexachlorophene. He then administered a snootful of 502A broth into each patient's nostrils. To gauge the effectiveness of the treatment, Steele compared the outcomes of twenty patients who received the 502A nasal spray with twenty who went through the same decontamination procedures and then received a squirt of sterile water. Six months later, thirty-two families remained in the study (some having moved away and others having taken additional antibiotics for unrelated conditions). Fifteen of seventeen families inoculated with 502A still had the strain firmly in place, and none had suffered a single recurrence of boils, rashes, or abscesses. By contrast, staph skin infections had returned to plague eleven of the fifteen families in the comparison group.[44]

Little Rock dermatologists lost their backup man and his unusual treatment in 1992, when Steele left to join the faculty of the Louisiana State University School of Medicine and practice at its affiliated children's hospital in New Orleans. A few years later, dermatologists were again knocking at his door, this time seeking help with nasty MRSA infections in schoolchildren and high school athletes. When Steele regaled them with stories of 502A, they seemed eager. But word quickly spread to hospital lawyers. "Next thing I knew, the administration was saying I would be putting the medical center in jeopardy," he recalls.

Admittedly, deliberately inoculating someone with *Staph. aureus*, even a benign strain, worries more than just lawyers. "I don't think there's a strain of staph on this planet that I would feel comfortable using in patients," says Columbia University microbiologist Frank Lowy, one of the world's leading experts on what makes this bacterium so dangerous. "We have to remember that staph has far too many potential weapons at its disposal. Even 502A has shown that it is capable of causing disease." A safer approach, Lowy argues, may lie in finding a potential competitor among less dangerous staph species, such as the ubiquitous skin residents *Staph. epidermidis* and *Staph. warneri.*

Around the same time that hospital lawyers in New Orleans were stifling Steele's plans to squirt staph into patients' noses, microbiologist Richard Hull, of Houston's Baylor College of Medicine, was having trouble believing the response to his wild idea of putting *E. coli* into patients' bladders. At the time, a visitor from the National Institutes of Health had come to Baylor interested in funding research projects that might benefit Americans with disabilities. Hull thought of the spinal-cord-injury patients at the neighboring Institute for Rehabilitation and Research, where he consulted. Many to most of these patients, paralyzed below the waist, suffered frequent bladder and kidney infections, some of them life-threatening. The problem was the indwelling urinary catheters they had to use, for the devices introduce bacteria into the urinary tract. Making matters worse, the repeated use of antibiotics to treat these infections eventually breeds multidrug resistance.

Hull considered how he might thwart such infections. Though the bladder normally remains sterile, some people carry bacteria in their bladders without ever getting sick, he recalled, and the presence of these benign bacteria appears to block the ability of other organisms to move in. And so Hull broached the idea of finding a relatively safe organism that he might deliberately introduce into the bladders of the paralyzed. "She asked me, 'What's the downside?'" Hull recalls of his discussions with the NIH grant supervisor. "I said, 'Well, on rare occasion it could cause serious infection, even death.' I figured that would be the end of it." Instead, to Hull's disbelief, she replied, "Okay, we can work with that." That was how desperate the need was for a solution to the paralyzed patients' highly drug-resistant infections.[45]

Hull had a specific bug in mind for his probiotic cure. Since the 1980s he had been corresponding with Catharina Svanborg, a gynecologist in Sweden. Svanborg had isolated a strain of *E. coli* that, year after year, showed up in the urine of a young girl whose urinary tract and kidneys remained in perfect health. "In the United States, a doctor would never know bacteria were there unless it caused an infection," Hull explains. "But in Sweden it's a regular part of preventive care to take a monthly urine sample."

Hull studied the bug, dubbed *E. coli* 83972, in laboratory cultures and combed through its genes to make sure that it didn't come equipped with the kind of molecular grappling hooks that enable some strains of *E. coli* to cause serious bladder damage.[46] Then, in two pilot studies in the late 1990s, urologists at the Houston rehabilitation center introduced *E. coli* 83972 into the bladders of fifty-seven paralyzed adults in their care. Before the probiotic treatment, all the patients had been plagued by recurrent urinary tract infections. A year later all but two remained infection-free. The two exceptions had both contracted a single, easily cured urinary tract infection caused by an organism other than their probiotic *E. coli*.[47]

Two Houston medical centers continue to use Hull's probiotic under an experimental protocol. Meanwhile, he and his students have begun testing a potentially more effective approach: placing urinary catheters in a solution of the probiotic and allowing a protective biofilm to form over the devices before they're used in patients.[48] Still more ambitiously, Hull would like to start using a genetically engineered version of *E. coli* 83972 that he created to eliminate any latent ability to cause problems. Specifically, he knocked out a gene that could potentially help the bacterium adhere to kidney tissue and another that might allow it to stick to the lining of the bladder.[49] In essence, Hull believes that the deletions will ensure that the bug will never "stick" where it doesn't belong, without compromising its ability to remain in the bladder in a free-floating state.

While tinkering with their creation in the lab, Hull's team couldn't resist going one step further: introducing a stripped-down, nontransferable resistance plasmid. The loop of resistance genes would enable their probiotic to survive the antibiotic treatments that the paralyzed frequently take for unrelated infections. Hull has not yet used his genetically enhanced strains in patients—a proposal still too audacious for his supervisors. But Hull is far from alone in wanting to build better bugs through engineering.

A SUPERHERO FOR THE MOUTH

In the summer of 1976, dental microbiologist Jeffrey Hillman was two months out of graduate school and newly hired at Boston's Forsyth Institute. One morning that summer, he glanced down at the petri dishes he had filled with tooth bacteria and saw two specks of red in a sea of white. The agar-filled plates contained a pH indicator that blanched white in the presence of acid—in this case the lactic acid produced by *Streptococcus mutans*, the main culprit in tooth decay. The red specks were colonies that had sustained damage to one of the genes controlling their production of this enamel-eroding chemical. The mutation hadn't slowed their growth in the least. That was hopeful news to Hillman, who went on to demonstrate that his acid-free mutants thrived on artificial teeth made of the biomineral hydroxyapatite—without causing cavities.[50]

"At the time, scientists all over the world were studying how humans become infected with *Strep. mutans*," Hillman recalls, "and whether or not we could replace the strain in someone's mouth with another." Research was showing, for example, that most of us inherit our *Strep. mutans* from our mothers and that some strains produce far more cavity-making acid than do others.[51] Moreover, once one strain of *Strep. mutans* takes up residence, it proves extremely difficult to dislodge it to make way for another. "We were trying all sorts of crazy things," Hillman says of the tactics he and his colleagues were using to eradicate *Strep. mutans* from the mouths of volunteers, before inoculating them with his experimental strains. "One time we were painting their teeth with iodine. Then we tried fitting their teeth with trays filled with antibiotics." But no matter how well Hillman and his colleagues beat back a person's original *Strep. mutans* or how quickly they recolonized a person's teeth with their acid-free version, the switchover never lasted for more than a couple of months. "Slowly but surely, a person's indigenous strain always came back," says Hillman.

By 1982, Hillman felt he'd exhausted every trick. That was when he hit on the idea of finding a bacterium that could do the hit job for him. Once he'd found this superaggressive strain, he reasoned, he could remove the gene that enabled it to produce acid. Hillman and two lab partners spent a year collecting saliva from Forsyth students and staff, ending up with hundreds of slightly different subspecies of *Strep. mutans*. They screened each bug for its ability to kill off other strains by growing them "cheek to jowl" in petri dishes. They knew they'd found their ideal candidate when they saw one pinprick colony clear a perfect circle in the lawn of other *Strep. mutans* around it. Analysis showed that the strain produced high levels of a novel bacteriocin, or natural antibiotic.[52]

Hillman and two lab partners became the first guinea pigs in 1985, painting the superstrain on their own teeth with cotton swabs. It immediately took up permanent residence in all three men, banishing their native *Strep. mutans* in the process.[53] A half dozen studies in rats confirmed their findings: once introduced, the bug consistently muscles out an animal's indigenous *Strep. mutans*. But Hillman's plan to simply knock out its ability to produce acid hit a glitch when the mutation proved lethal. Some strains of *Strep. mutans*—including this one—use lactic acid as a way to dispose of metabolic waste that would otherwise build to toxic levels.

Hillman solved the problem by knocking *in* an extra copy of a gene for alcohol production, for this allowed the bug to divert its metabolic wastes in a different direction. "The strain we ended up with was indistinguishable from the original except for two genetic modifications that we could spell out letter by letter," says Hillman. Studies with rats showed that the novel *Strep. mutans* kept teeth virtually cavity-free on a high-sugar diet that would normally destroy them. Importantly for potential safety concerns, Hillman showed that the original strain residing in his mouth and those of his partners had not spread to any family member over the course of more than a decade.[54]

In 1998 Hillman gathered up the results of his research and approached the FDA for permission to use his genetically engineered *Strep. mutans* in human volunteers. "Fortunately, I had no idea what lay ahead," he says. First, FDA regulators asked Hillman to cripple his bug to ensure that it could be removed should it cause problems. "When we asked them what kind of problems, they had no idea," he recalls. "I guess we were setting a precedent for the evaluation of genetically altered organisms."

Hillman knocked out more genes, this time rendering his microbe unable to survive without a twice-daily supplement of an amino acid rarely found in food. To keep the bacteria alive, volunteers would need to swish with a mouthwash containing this nutrient on a daily basis. "Hopefully, once we've shown that it's safe, they will allow us to use a fully functional organism," he says.

Confident that he'd satisfied all possible safety concerns, Hillman reconvened with his FDA reviewers in March of 2004, having formed a biotech company (Oragenics) to finance the necessary clinical trials. To his surprise, he learned that his *Strep. mutans* had been lumped into the same category as potential bioweapons. The review committee set forward its requirements: Hillman could begin with a small safety trial of ten people, but they all had to be toothless—that is, full denture wearers whose false teeth could be dropped into bleach to expunge the organism after a week of testing. As recruitment got under way, the FDA added a few other rules: the volunteers could not have children in their homes; their spouses had to be full denture wearers; and both they and their spouses had to be robustly healthy and under age fifty-five. "We screened more than a thousand potential volunteers and found two who met all requirements," says Hillman. The miniature two-person trial proceeded without a hitch in 2006, with no adverse side effects and complete elimination of the organism at the end of seven days.

As of mid-2007, Hillman was still awaiting approval for a study with people who have their own teeth. But while the FDA gives the go-slow to the first man to create "designer microflora" for human use, scientific journals are filling with the early experiments of those eager to go further. Mixing and matching genes from different organisms, microbiologists around the world are creating transgenic laboratory bugs with the potential to do much more than crowd out their problematic counterparts in the human body. Already, one European group has safety-tested their transgenic bacterium in ten patients—while sequestering them in a hospital containment ward kept under negative pressure to prevent its accidental release.

TRANSGENIC PROBIOTICS

The forty-three-year-old Dutch farmer was packing his bag to leave the hospital when the nurses caught him. "He felt so much better after three days, he was ready to go home," recalls Maikel Peppelenbosch, a Belgian molecular biologist working at the University Medical Center Groningen, in the Netherlands. "We had to explain to him that he was not free to leave no matter how wonderful he felt." Three days earlier, on a spring morning in 2003, the middle-aged farmer had swallowed his first twice-daily handful of ten small capsules, each filled with some 10 billion cells of the cheesemaking bacterium *Lactococcus lactis.* That small act entered the Dutchman into the history books as the first human deliberately colonized with transgenic bacteria. The live bugs he'd swallowed carried and expressed the human gene for the immune-calming cytokine interleukin-10.[55]

Researchers have long known that lab animals that can't produce interleukin-10 develop severe inflammatory bowel disorders similar to the Crohn's disease that had debilitated the Dutch farmer for more than twenty years. The immune system of an IL-10-deficient mouse, like that of a Crohn's patient, loses its tolerance to the digestive tract's normal microflora. The result: excruciating and sometimes life-threatening inflammation and intestinal ulcers. But efforts to administer interleukin-10 as a therapeutic drug are fraught with problems. It's difficult to get enough of the immune-calming chemical into the bowel, where it's needed, and even trickier to keep it out of the rest of the body, where too much can cause dangerous immune suppression.

In 1999 the Belgian molecular biologist Lothar Steidler came up with a novel solution. He took the human gene for IL-10 production and knocked it into the chromosome of *L. lactis,* a live-culture bacterium that lingers for twelve to twenty-four hours in the digestive tract before being swept away in a person's stool. That's long enough for a twice-a-day supplement to deliver the calming cytokine to intestinal tissues without risking bodywide immune suppression.

In the same year, Steidler also succeeded in using his transgenic to cure the mouse equivalent of Crohn's disease.[56] But at the time, he was just one of many young scientists who were concocting laboratory strains of transgenic bacteria with the hope that these bugs might someday deliver drugs or vaccine antigens to human patients. Safety concerns demanded that all these researchers keep their genetically engineered "Frankenbugs" under strict biological containment—along with any colonized lab animal. Like the FDA, European health agencies had only begun to grapple with the potential consequences of using genetically modified organisms in people. A bacterium that produced a powerful immune suppressant such as IL-10 posed a particular danger. Even if the transgenic itself proved harmless, disaster could result if it transferred its new gene to disease-causing microbes, giving them the power to switch off any infection-fighting immune response.

Then again, Steidler appears to have been more clever than most. When he endowed his *L. lactis* probiotic with the ability to produce interleukin-10, he knocked the human gene dead center into an existing gene that the bacterium needs to make the nutrient thymidine. So, like Hillman, Steidler ended up with a designer bug that couldn't survive long unless deliberately fed. Also in Steidler's favor, *L. lactis* was not a normal member of the body's microflora. When consumed in dairy products, ordinary *L. lactis* disappears from a person's intestinal tract within a day or two. Finally, Steidler's genetic engineering ensured that even if a transgenic *L. lactis* did share its IL-10 gene with other microbes, the only place the recipients could integrate the new DNA would be in the middle of their thymidine-producing genes. So they, too, would become nutritional cripples.

Among those most impressed with Steidler's genetic sleight of hand was Peppelenbosch. The two young Belgians were working at the same lab bench, completing their separate postdoctoral research projects at the Flanders Institute for Biotechnology, in Ghent. Steidler subsequently left for Ireland, taking a professorship at University College Cork. Peppelenbosch left for the Netherlands, where he headed his own laboratory within the University Medical Center Groningen. "When I learned Lothar was not getting his clinical trials off the grounds in Ireland, I offered to give it a go with the Dutch regulatory authorities," Peppelenbosch says.

"Fortunately, Lothar designed this bacterium very well," he adds. "Getting regulatory approval was a lengthy process, but we received not one formal complaint." Within eight months of approaching Dutch health authorities, Peppelenbosch had approval to conduct a safety trial with ten Crohn's patients, none of whom had responded to conventional treatments such as steroids. "These were patients for whom taking out the bowels was their last remaining option," he says. Peppelenbosch's funding for the historic trial came by way of the United States and a private research grant from the billionaire businessman Eli Broad, whose son has long suffered from the disease.

And so, over the course of two and a half years, doctors at Amsterdam's Academic Medical Centre cycled ten patients through its one-room biohazard isolation ward. Like the forty-three-year-old farmer who inaugurated the study, most of the patients reported a dramatic reduction in their symptoms, says Peppelenbosch. "But this wasn't to show efficacy, only safety," he quickly adds. "We did not even use a placebo for comparison." The researchers had judged it unfair to ask the desperately ill volunteers to risk getting a dummy treatment on top of the hardship of living in isolation for twelve days.

As a safety test, however, the trial proved a full success.[57] The transgenic *L. lactis* produced no ill effects, and completely disappeared from the volunteers' stool within twenty-four hours of their swallowing the last capsule in their weeklong course. As expected, the volunteers' symptoms returned within a few weeks of their returning home, prompting several to plead to continue the unconventional treatment. "We couldn't, of course," Peppelenbosch explains. "The regulatory authorities are right. We have to be careful." Still, he's hopeful that the ten patients, along with fifty more, will be allowed to participate in a planned follow-up trial in the summer of 2008. "Now that the safety trial showed that biological containment worked," he says, "we expect that government regulators will allow us to do the next trial on an outpatient basis." As of mid-2007, the researchers were still awaiting word.

Meanwhile, Steidler and Peppelenbosch are working on an even more targeted deployment of transgenic probiotics. Their idea is to give various types of drug-producing bacteria additional genes for producing antibodies that cause the modified bugs to adhere to specific tissues in the body. For example, a cancer-fighting probiotic might be endowed with a gene for producing antibodies that make it stick to the surface of tumor cells.

At the same time, in laboratories around the world, scientists are testing dozens of other transgenic probiotics in animals. These include several different kinds of vaginal bacteria that secrete HIV-killing drugs. Canada's Gregor Reid, for example, has teamed with researchers in the United States and Australia to enhance his vaginal probiotic *Lactobacillus reuteri* with human and recombinant genes that produce a cocktail of three proteins that block the ability of the AIDS virus to bind to, fuse with, and enter the immune cells it normally destroys.[58] Closer to clinical trials is an HIV-fighting lactobacillus now in development at the biotherapeutics company Osel, in Santa Clara, California. The idea behind this creation dates back to the mid-1990s and a quiet moment in the laboratory of the Stanford University physician-scientist Peter Lee. Lee recalls daydreaming of ways to block viruses from entering the body. "What I came up with," he says, "was the idea of trying to harness the bacteria that live on our mucous membranes."[59] For viruses almost always enter the body through these moist, semiporous membranes.

For nearly a year Lee scoured the scientific literature and interrogated colleagues in an effort to turn up related research. To his surprise, he says, he found none. "But the more I thought about it, the more it made sense to me. If our resident bacteria already form a protective barrier, why not engineer them to fight viruses more effectively?"

What Lee did find was the published work of University of Pittsburgh research gynecologist Sharon Hillier. Hillier was working with colleagues in Africa to develop inexpensive ways women could protect themselves from HIV in communities where condoms were too costly or culturally unacceptable. For a start, she had found that women with a healthy vaginal flora—that is, an abundance of lactobacilli—had *half* the HIV infection rate as did women who lacked these protective bacteria.[60] She had also begun screening strains and species of lactobacillus to find those that might prove most protective. The best candidates turned out to be those that both produced a high level of virus-inhibiting hydrogen peroxide and formed a natural biofilm that "glued" their cells to the vagina's surface.

Lee began talking with Hillier about the idea of enhancing her HIV-fighting vaginal bugs with genetic engineering. And in 1998 he founded Osel to develop his dream. Lee's scientific team at Osel enhanced the vaginal bacterium *Lactobacillus jensenii* with a human gene for the cell protein CD4—HIV's molecular target. When mixed with cultures of human cells, the transgenic bug completely blocked infection with a laboratory strain of HIV, and it reduced by half the infectiousness of a strain isolated from a patient.[61] In 2006 the team announced they had built what they thought would be an even stronger HIV-fighter: a transgenic *L. jensenii* that produced a virus-destroying protein known to prevent HIV infection in monkeys.[62] The gene for this protein—cyanovirin N—came from the sparkling blue cyanobacterium *Nostoc ellipsosporum*.[63] In early 2007 Osel's scientific team was testing the ability of their new HIV-blocking bug to block HIV infection in animals. Lee, meanwhile, continues to divide his time between Osel and his research at Stanford University, where he now dreams of building microbes that will block the viruses that cause leukemia.

While Hillman and Lee work at the forefront of efforts to genetically enhance our native microflora, others are tinkering with the possibility of creating live transgenic vaccines. These futuristic vaccines would consist of harmless members of our microflora engineered to express the antigens, or molecular markers, of dangerous pathogens. In theory, once such a bug took up residence in our bodies, it would induce our immune systems to produce antibodies against the bad guy whose black hat it wore.

Among the first to produce a working example of such a live transgenic vaccine is Rockefeller University's Vincent Fischetti. In 1995 Fischetti created a strain of the mouth bacterium *Streptococcus gordonii* that studs its surface with the signature antigens of *Strep. pyogenes*.[64] Similarly, biologists at the State University of New York have created a transgenic strain of *Strep. gordonii* that wears the antigens of *Porphyromonas gingivalis*, the culprit in bone-destroying periodontal disease.[65] And at France's Pasteur Institute, immunologists have created a number of live vaccines out of the yogurt bacterium *Lactobacillus plantarum*, including one that produces a segment of the tetanus toxin and another that vaccinates against *Helicobacter pylori*, the stomachulcer bug.[66] Other scientists have created transgenic bacteria that immunize against cholera, salmonella, shigella, listeria, tuberculosis, plague, anthrax, and even cancerous tumors—all of which have produced modest results in tests with lab animals.[67]

Still other researchers are building transgenic bacteria that prompt the immune system to produce antibodies against problematic chemicals of the body's own making. Swiss scientists, for example, have developed a transgenic version of the yogurt bacterium *Lactobacillus johnsonii* that wears the human version of the allergy-inducing IgE antibody on its surface. When injected into animals, the bug induces their immune systems to produce a different kind of antibody (IgG) that mops up the problematic IgE.[68]

"The use of live bacteria carriers constitutes a powerful tool to achieve efficient delivery of vaccine antigens," German immunologist Eva Medina wrote in a 2001 review of the new and rapidly growing field. "Almost unlimited possibilities are offered for the exploitation of this system."[69]

Not everyone is so enthusiastic. "The prospect that genetic modification might improve probiotic microbes must be seriously balanced against the potential of turning harmless, beneficial microbes into dangerous pathogens," argues geneticist Joe Cummins, recently retired from the University of Western Ontario. Among the greatest dangers, says Cummins, is that the bacteria introduced into one person can easily move to another. When it comes to vaccinating against dangerous bacteria, such independent spread might prove hugely beneficial. But it makes assurance of safety more difficult because the live vaccines would be reaching the immune-compromised as well as the healthy.[70]

Cummins cites another danger: a live vaccine that takes up residence in the mouth, throat, or gut could have the unintended effect of inducing tolerance instead of a disease-fighting response. "Once a bacterium becomes a permanent part of the body's ecology, the immune system will likely begin recognizing it as self," he argues, "and in that state it will no longer induce antibodies against disease-causing bacteria." Cummins refers to recent findings by immunologists who have shown that the repeated presentation of almost any antigen turns *off* the body's immune response, at least if it occurs in the absence of danger signals such as tissue injury. Presumably, this is how the body develops tolerance to the antigens in our food as well as the microflora of our digestive and upper respiratory tracts.

Cummins takes particular issue with live vaccines aimed at the immune system's own signaling molecules and antibodies. "Experience tells us that interfering with the immune system can lead to nasty surprises," he warns. As an example he cites the recent case of a harmless mousepox virus that unexpectedly turned deadly when Australian researchers gave it a gene for a protein found on mouse egg cells. The scientists were trying to create a mouse contraceptive, but vaccinating mice with the recombinant virus wiped out an entire arm of their immune systems.[71]

PROBIOTICS FOR LIVESTOCK

Just as a new generation of vaccines and probiotics can take human medicine in a fresh direction, they likewise offer a way to take our livestock off the antibiotic treadmill. In the spring of 1998, the Agricultural Research Service's "Food and Feed" team, in College Station, Texas, seemed to have scored the proverbial home run. The lab had produced an easy-to-apply probiotic spray that prevented newly hatched chicks from being colonized by salmonella, the dangerous food bacterium that frequently contaminates undercooked eggs and chicken. The spray, which farmers could mist over hundreds of chicks in minutes, contained a blend of twenty-nine kinds of harmless and tenacious bacteria that David Nisbet and his ARS colleagues had isolated from the digestive tracts of healthy hens. A single treatment conferred lifelong protection against salmonella in more than 99 percent of chicks and proved comparable to growth-promoting antibiotics in fostering their rapid weight gain.[72]

The researchers dubbed their product Preempt, and its FDA approval in March 1998 inspired national news headlines ("Bacteria Banish Fowl Bugs") and video reels of farmers happily spraying flocks of adorable yellow peepers. In its press releases MSBioscience, the poultry supplier licensed to market Preempt, described it as "the first of a new generation of microbial products designed to enhance the body's defenses against disease." In speaking to reporters, codeveloper John DeLoach added that the spray's widespread use might someday reduce the dangerous pathogens in eggs and raw chicken to levels "so low that they aren't important."[73]

But the media celebration soon faded, and an initial flood of demand for Preempt dwindled to a few long-standing orders from organic farmers. In 2002 MSBioscience quietly took its product off the North American market. What happened? Microbiologist Todd Callaway, who joined the USDA Agricultural Research Service's "probiotics" team in 1999, says, "The product was a success, but it just wasn't as cheap as antibiotics. Preempt cost around a penny a bird to use, and farmers could get the same growth-promoting effect with antibiotics for a third of a cent." As for Preempt's salmonella-slashing benefits, salmonella doesn't sicken chickens as it does people, and no one was offering to pay

Preempt-using farmers more for delivering chicken and eggs free of the dangerous food pathogen.[74]

Far from discouraged, the Agricultural Research Service team has come back with a piggy version of Preempt that promises to reduce salmonella and other dangerous human pathogens in pork.[75] But this time around, the researchers aimed for a product that, first and foremost, would make money for its farmers, with reductions in people-infecting bacteria a "free of charge" bonus for consumers. They came up with a probiotic that protects against what may be the greatest bane of modern pig farming: scours, a diarrhea-inducing infection produced by a dozen-odd strains of toxic *E. coli*. The *E. coli* strains that cause pig scours pose no danger to consumers, but they kill millions of piglets each year.[76] The young animals often develop the infection during the stressful time right after they're taken from their mothers and crowded into nursery pens.

"It's a sad thing to see," says Francis Forst, who supervises the rearing of some quarter-million pigs a year at the Missouri Sow Center, in Lamar. "Piglets with scours don't eat much, don't grow, they just kind of lie there, rough-haired and chalky with their rear ends raw and red, like babies with the worst kind of diaper rash." During outbreaks it's not unusual for a farmer to lose 20 percent of piglets, Forst explains, and the survivors tend to grow slowly and never reach full weight. "No antibiotic works well against it," he adds. In fact, antibiotics tend to make toxic *E. coli* infections even more deadly.[77]

Eager to participate in the Agricultural Research Service's field trial, Forst found that his workers could easily inoculate more than 750 freshly weaned piglets a day by squirting the probiotic into the animals' mouths from a calibrated pump bottle. In addition to preventing scours, he says, the formula cured many of his already infected piglets. "The change was dramatic," he recalls. "You could see those pigs going from chalky white back to a nice pink, up and jumping around, and filling out nice and plump."

The net result for the Missouri Sow Center's profit margin: a 2.5 percent decrease in mortality. That may not sound like much, but Forst multiplies it by the extra $50 his operation brings in for every additional pig delivered to market. During the year Forst participated in the probiotic field trial, the total came to around $312,000.

Next the ARS team hopes to test the probiotics in newborn piglets, many of which likewise sicken with scours. "Inoculating the pigs within the first twenty-four hours of life also increases our chances of permanently colonizing them with the right kinds of bacteria," says study leader Roger Harvey. Pig farmers such as Forst like the idea, as it would be easy to add the probiotic to the schedule of "baby shots" and mineral supplements that the newborn piglets receive. "We're already picking them up," Forst says. "It would be easy enough to add a squirt of Roger's product."

Meanwhile, MSBioscience is finding new interest for Preempt in Europe, where agricultural antibiotics have been banned. "Whether you're working with chicks, piglets, or calves, it's always going to be to your advantage to get the good bugs in place before the bad ones can take up residence," Callaway concludes. "But so long as there's an antibiotic available that can get an animal to market for less money, you can't blame the farmer for wanting the cheaper product."

A SECOND NEOLITHIC REVOLUTION

Ever since Semmelweis and Lister began sterilizing their hands and surgical instruments, the microbes in our environment have been viewed almost uniformly as threats to our health. Today we have chlorine in our drinking water and germicides in our cleaning products. Simply preparing a meal has become an exercise in sterilizing food and kitchen surfaces. Besides depriving us of the immune-calming benefits of life's harmless bacteria, our attempts at disinfecting our lives have resulted in only partial victory in our fight against infectious disease.

"Whenever you make a sterile surface, you become a victim of whatever falls on it," argues Rockefeller University microbiologist David Thaler. "It's like plowing a field and not planting anything but instead trying to live on whatever weeds happen to pop up."[78]

Thaler is best known, in the scientific community, for his ideas about the genetic changes involved in cell development and species evolution.[79] In recent years, however, his intellectual explorations have focused on what he sees as the *next* great leap in the advancement of our own species: an overhaul in our dealings with the bacterial world.

He calls it the dawning of a "microbial Neolithic revolution." For just as Stone Age hunter-gatherers transitioned to modern humans who cultivate their food, Thaler sees a second Neolithic revolution in which we exchange our microbe-hunting ways for methods aimed at domesticating bacteria and deliberately seeding our environment with carefully selected strains and species.

"In this future we will not sterilize external surfaces such as skin, air, and doorknobs," Thaler says. "The folly of doing so will be recognized as equivalent to that of sterilizing the intestines today." The goal of "purity" will no longer be "99.9 percent germ-free," he adds. Instead, it will lie in achieving the perfect balance of beneficial microbes. We will no longer douse our hands, faces, and bodies with antibacterial soaps; we will wash them with probiotic mixtures shown to enhance health. Laundry detergents will contain sporulated or freeze-dried bacteria that "awaken" to keep clothes fresh. We will buy shoe inserts infused with organisms isolated from the world's freshest-smelling feet. Instead of futilely trying to disinfect public bathrooms, cleaning crews will spray toilets and doorknobs with tenaciously territorial "good" bugs. Subway straphangers will grab onto handholds impregnated with bacteria that kill cold and flu viruses on contact. Hospitals will finally eradicate their plagues of deadly, drug-resistant microbes with strains of bacteria that enhance their patients' immune systems and *help* them fight disease.

Standing between our present and Thaler's imagined future is a huge scientific effort. "For a start, we need methods to inventory our microbiological environment, both intimate and global," he says. As a jump-start toward such a future, the last decade has brought many great advances: we have developed a more complete census of our body's normal microflora, and we have deepened our understanding of the key character differences between our best-behaved tenants and the small handful of chronic troublemakers. We're also beginning to understand what enables one strain or species of bacteria to crowd out a competitor.

But for the most part, we've studied our bacterial friends and foes only one or two at a time rather than as the dynamic and diverse communities that are their natural state. We now know, for example, what mix of bacteria predominates in a sweet-smelling mouth, a disease-resistant throat, and the digestive tract of a healthy person, pig, or chicken. Are we ready to take the next step—to determine the perfect microbial mélange to render, say, a bandage impervious to *Staph. aureus* or a cutting board inhospitable to salmonella?

Even after we've characterized the breadth of our microbial environment, the task will not be over. Like any ecosystem, microbial communities evolve and suffer disruption. Thaler envisions a future in which we continually monitor these invisible dynamics. In the outside environment, this monitoring could be an extension of the technology currently in development for detecting bioterrorist threats. On a more personal level, Thaler envisions doctors approaching a physical exam with the equivalent of a high-powered microscope, ensuring that all the right microbes are in the right place and in optimal balance. "It's also worth asking," he says, "what would it take to recognize and nurture beneficial incarnations of Typhoid Mary." By this he refers to people who not only benefit from their healthful microflora but spread it to others.

Circling back to food and drink, where we humans took our first baby steps in domesticating microbes, scientists are already asking whether it's time to go beyond cultured yogurts and cheeses to add back some of the microbial life that once infused the meat that we hung from our rafters and the produce and grains we crudely stored in cellars and cisterns.

"The idea of deliberately putting bacteria back into meat or produce makes the people in the Food Safety and Inspection Service roll their eyes in horror," says Agricultural Research Service microbiologist Todd Callaway. "But the truth is, nature doesn't abhor a vacuum, it *loves* a vacuum. It gives the bad guys a chance to do something." If his colleagues can get past their taboos, Callaway says, there may be a useful place for such products as meat and produce sprays containing "nonspoilage organisms" such as acid-producing lactobacilli.

"Humans live in a world that's also alive," Thaler concludes. "Humans don't function optimally when they attempt to embed themselves in a sterile environment. We would do better by learning to live up close and personal in a seamlessly continuous living world."

PART VII

FIXING THE PATIENT

[The body's] arsenals for fighting off bacteria are so powerful and involve so many different defense mechanisms that we are in more danger from them than from the invaders. We live in the midst of explosive devices; we are mined.

—physician/immunologist LEWIS THOMAS, 1978

THE DRAGON WITHIN

To this day, the smell of overripe pumpkin takes Kevin Tracey back to an evening in May 1985, when a badly burned little girl arrived in the emergency room at New York Hospital. At the time, Tracey was a second-year resident in training as a neurosurgeon. Eleven-month-old Janice was scalded over 75 percent of her body. Minutes earlier, she had crawled between the legs of her grandmother just as the woman turned from the stove with a pot of boiling water. As Tracey finished examining the critically injured baby, nurses slathered Janice's body with a thick layer of Silvadene antibiotic cream. Its acrid, sweet odor became Tracey's sensory snapshot of the night and the tumultuous month that followed.[1]

Tracey knew that Janice's chances of survival were slim, for even the thickest layer of antibiotic ointment was no substitute for her lost skin—her body's barrier to the invisible sea of bacteria that surrounded it. Once these bacteria infiltrate beyond skin and mucous membranes, the result can be bodywide inflammation, or sepsis, a little-understood condition that, in the 1980s, was thought to be caused by bacterial toxins. Sepsis frequently turns deadly when it takes one of two severe forms. Janice would suffer both.

Four days into Janice's hospital stay, her blood pressure plummeted toward zero as her blood vessels suddenly became so leaky that they lost their fluid into surrounding tissues. She was in septic shock. For twelve hours Tracey and a dozen other doctors and nurses worked to keep the little girl alive with blood-vessel-constricting drugs and massive infusions of intravenous fluids. All the while, they knew that if her shock persisted for more than half a day, their efforts would prove inadequate to save her organs and extremities. They just didn't know if or when the condition would resolve.

At 9:00 a.m. the next morning, Janice's blood pressure rose and stabilized almost as quickly as it had collapsed. She lost no fingers or toes, but likely sustained damage to her lungs, kidneys, and other organs. Over the following three weeks, Janice's body slowly recovered from the crisis, and on her first birthday her family gathered in her hospital room to celebrate with balloons and chocolate cake.

The next day, as Janice finished a lunchtime bottle in the arms of a nurse, her eyes rolled back in her head and her heart monitor went flat. Tracey recalls the surrealistic experience of responding to the nurse's shouts, administering CPR, and assisting a cardiac surgeon as they tried, in vain, to restart the little girl's heart. Having survived the lightning bolt of septic shock, baby Janice ultimately succumbed to organ failure from its stealthier cousin, severe sepsis.

But why? Then, as today, no one understood why severe sepsis or septic shock developed in some patients but not in others. Sometimes the obvious answer was the presence of an overwhelming bacterial infection. In other cases, such as baby Janice's, the bacteria driving the conditions remain so scant and hidden that they never show up in tests of blood and urine or upon the dissection of organs at autopsy.

"I found the experience so frustrating and perplexing that it changed the course of my career," says Tracey. "Instead of becoming a neurosurgeon, I chose research into sepsis." Though death certificates seldom cite it by name, sepsis kills around a quarter million Americans each year, often as a "complication" of severe injuries, surgery, or a chronic disease.[2] "Whenever you hear of someone dying of a bacterial infection in this day and age," says Tracey, "the real cause is almost always sepsis."

Among those at highest risk are the seriously burned, the immune-compromised, diabetics, and the paralyzed. In the latter two groups, poor circulation and open sores predispose to bacterial blood infections. The paralyzed actor Christopher Reeve, for example, died from sepsis secondary to an infected pressure sore, a common problem among the bedridden and wheelchair-bound. Experts estimate that millions more succumb to sepsis each year in the final stages of a terminal illness such as cancer or heart disease.

And while sepsis can be triggered by the presence of any sort of bacteria in forbidden territory, some kinds are more associated with it than others. Among the most infamous is the virulent USA300 strain of MRSA that killed twenty-one-year-old college football star Ricky Lannetti in 2003 and forced the leg amputations of the young Baltimore women treated for MRSA pneumonia at Johns Hopkins the same winter; all three were victims of septic shock. The "toxic shock" streptococcus that killed Muppeteer Jim Henson in 1990 is likewise known for its tendency to trigger sepsis.

What Tracey's twenty-five years of research have helped reveal is that it's not *bacteria* that wreak the deadly damage of sepsis, at least not directly. In the fall after Janice's death, Tracey teamed up with immunologists Anthony Cerami and Bruce Beutler at Rockefeller University (conveniently next door to New York Hospital's burn unit). Cerami had been working with an immune-system-signaling molecule, or cytokine, that caused severe weight loss, or wasting disease, in patients with chronic infections or late-stage cancer. The molecule had become known as tumor necrosis factor (TNF) because it was the primary weapon deployed by cancer-killing immune cells such as macrophages. As it turned out, macrophages likewise deployed TNF to kill bacteria and infected cells. In a side project, Beutler had decided to see what would happen if he injected mice with TNF-blocking antibodies right before he injected them with a shock-inducing dose of "endotoxin," a molecule found on the cell surface of 90 percent of all bacteria. All the mice survived.[3] This small study raised a major scientific question: Could it be that TNF, not bacterial toxin, was the poison behind deadly sepsis?

The answer was yes, and Tracey provided the proof: First he showed that injecting lab animals with nothing but TNF could cause septic shock.[4] He then demonstrated that he could prevent shock by giving animals TNF-blocking antibodies before introducing bacteria into their bloodstreams.[5] The implication turned the medical world's understanding of sepsis on its head: *it was a person's own immune system, not bacteria, that produced deadly sepsis.*

So-called endotoxin, for example, turns out to be an inert chemical, lipopolysaccharide. Its presence on the surface of most bacterial cells serves as a kind of bar code that alerts the immune system to their presence in the body. We now understand that bacteria-fighting immune cells such as macrophages and neutrophils react to this red alert by deploying TNF and other biochemical weapons. Trouble starts when the bacteria persist despite the immune system's early efforts to destroy them.

Their stubborn presence, even in small amounts, can whip the immune system into an ultimately self-destructive frenzy. The same inflammation that, when focused, helps speed immune cells and antibodies to the specific site of an infection becomes deadly when it storms through the entire body, opening blood vessels indiscriminately until circulation collapses and organs and extremities begin dying for lack of oxygen.

Tracey's landmark papers appeared in the top-tier journals *Science* and *Nature* in 1986 and 1987. By the early 1990s, biotech companies were rushing anti-TNF therapies into patient trials. They failed spectacularly. "Trying to block TNF after the onset of shock is like closing the barn door after the horse has bolted," Tracey says with hindsight. "You can't reverse the damage once high levels have been released." Given the complexities of the body's immune response, Tracey and others also suspected that many molecules besides TNF played crucial roles in different patients and at different points in the rapid cascade of septic shock. While researchers continue to plumb these mysteries, the wreckage of start-up companies that lived and died by potential sepsis cures has bred wariness among the investors who are needed to bring new treatments to market.

For his part, Tracey shifted his interest to methods for treating the less flamboyant form of sepsis that ultimately killed baby Janice. He suspected that the slow organ failure of severe sepsis stemmed from a gradual bodywide breakdown of normal barriers. "In severe sepsis, while you don't see the major organ damage associated with septic shock, you have cells leaking fluids, bile mixing with blood, oxygen and water mixed together in the lungs," he explains. "It may be that when these barriers begin to fail at the cellular level, it's not long before the organs themselves give out." Patients and lab animals with severe sepsis have relatively nontoxic amounts of TNF in their bloodstream, he adds, certainly not the high levels associated with acute shock. And if anything, treating severe sepsis with anti-TNF drugs makes matters worse.

In 1994 Tracey began investigating severe sepsis in earnest, having become a senior scientist at Feinstein Institute for Medical Research, in Manhasset, New York. He and his students searched for cytokines other than TNF that might be released during the syndrome's slower unfolding. Of particular interest, they isolated a protein known as high mobility group box 1, or HMGB1—a seemingly poor candidate, as it can be found in virtually all healthy cells. But further tests showed that patients who died of severe sepsis had wildly high levels of HMGB1 in their blood. Tracey's lab team also demonstrated that blocking HMGB1 with monoclonal antibodies pulled lab animals back from even advanced stages of the disease.[6]

Tracey and many immunologists now see severe sepsis and septic shock as two distinct syndromes rising from the same "frustration" on the part of the immune system. When macrophages and other bacteria-killing cells fail to eliminate bacteria early in an infection, they can begin overexpressing their biochemical call to arms. If that overexpression is in the form of TNF, the rapid result can be septic shock. If it is HMGB1, the immune system initiates the slower dissolution of severe sepsis.

Given that severe sepsis unfolds more slowly than does septic shock, anti-HMGB1 therapies just might succeed where anti-TNF treatments failed, says Tracey. While it's now clear that anti-TNF treatments have to be administered *before* the free fall of septic shock begins, HMGB1 antibodies save lab animals when given up to thirty-six hours after the onset of severe sepsis. In 2007 the Maryland biopharmaceutical company MedImmune and Tracey's own company, Critical Therapeutics, were collaborating to bring anti-HMGB1 antibodies into patient trials.

Tracey's interest shifted again in 2000, this time to an earlier window of opportunity for intervention—one that might stop the immune system from going down the path toward *either* malfunction. He and his students knew that the vagus nerve, which controls vital functions such as breathing and heart rate, also modulates inflammation. They found that by electrically stimulating the nerve, they could save septic lab animals.[7] Significantly, vagus nerve stimulation did not block the immune system's ability to fight bacteria. It just kept it from flipping into self-destructive meltdown. Over the next four years, Tracey and his colleagues teased out precisely how the vagus nerve's release of acetyl-choline—the same neurotransmitter that regulates organ function—engages immune cells in a way that dials back their overproduction of inflammatory cytokines such as TNF and HMGB1.[8]

"It may be that we can develop a pacemaker-like device for patients that blocks TNF and HMGB1," he says of the potential applications of this new understanding. Similarly, Tracey's colleagues at the Feinstein

Institute have shown they can curb sepsis in mice by stimulating the vagus nerve with drugs or by mimicking its neurotransmitter, acetylcholine, with small amounts of nicotine.[9] Therapies based on these findings remain a long way off, especially given that biotech companies have become wary of investing in any kind of antisepsis treatment. Even so, immunologists such as Tracey express great hope that eventually they'll produce cures. "We have to remember that in the early 1980s, there was nothing you could give a septic animal that could save it," he notes. "Now we have lots of ways to save septic animals. History supports the likelihood that, if you really understand how to save an animal, you can eventually save patients."

More profoundly, says Tracey, our new understanding of what causes sepsis has helped broaden infection-fighting strategies beyond an exclusive focus on killing microbes, to include treatments that can correct the body's response to their trespass. "The ancient Greeks were right," he says. "For two thousand years their theory of medicine focused on balancing humors, on trusting that if the body was in balance, the patient would recover. That all changed in the nineteenth century, with the germ theory of disease." While germ theory gave us many lifesaving medicines, the discovery that the immune system can itself become toxic has brought the focus back to fixing the patient.

While the world waits for antisepsis cures, critical-care doctors are keeping more patients alive through the crisis of septic shock with advances in supportive care such as the rapid infusion of replacement fluids, better vessel-constricting drugs, and even tight control of blood glucose levels.[10] Today nearly two-thirds of sepsis victims survive, up from barely half some twenty-five years ago. Alarmingly, this hasn't been enough to counter the rising toll. The Centers for Disease Control recently estimated that the annual incidence of sepsis in the United States has risen from around 164,000 in 1979 to nearly 660,000 in 2000—or from 83 cases to more than 240 cases per 100,000 Americans.[11] Worse, this increase turned out to be a gross underestimate, for it counted only patients whose laboratory cultures confirmed the presence of bacteria in their blood. But as critical-care doctors know too well, the infections underlying sepsis often remain undetectable.

Studies by the University of Pittsburgh's Derek Angus showed that the total number of sepsis cases—regardless of positive bacterial cultures—totaled nearer the million mark in 1999 and were rising at a rate of more than 6 percent a year.[12] "The incessant growth of the severe sepsis epidemic in this country ought to be cause for grave alarm," Angus warned. In part the increase stems from a general aging of the population, with the elderly being more at risk for sepsis than those under eighty-five. But the last fifty years has also brought a change to the "typical" human body.

ENHANCING THE BIONIC HUMAN

Today U.S. surgeons implant tens of millions of artificial parts every year, from pins in broken bones and stents in blood vessels to joint replacements and mechanical heart valves. And every year over 1.5 million of them become infected.[13]

Early in the history of medical implantation, surgeons thought it enough to thoroughly sterilize the devices they installed and put their patients on antibiotics for a day or two as added insurance. But the ranks of the bionically enhanced had not reached the million mark, in the 1980s, before it became obvious that our bloodstreams were not the microbially sterile places we had assumed them to be.

We now know that the bacterial residents of skin and mucous membranes regularly stray into the internal recesses of the body, not just through gaping wounds but also through life's ubiquitous pimples, inflamed gums, and the constant insults of solid food passing through the digestive tract. This presents no problem to healthy living tissues, which can quickly summon an immune response. By contrast, steel, Plexiglas, medical textiles, and the like offer invitingly inert territory where bacteria can linger awhile, perhaps raise a family, or better yet build the elaborate citadels known as biofilms.

It turns out that, given the chance, most bacteria shun the solitary and drifting lifestyle that scientists glimpse in test-tube cultures of a single species. In nature they readily organize themselves into diverse communities, divvying up duties from food manufacture to garbage disposal to public defense. The resulting biofilms are naturally resistant to antibiotics—in part because of their ability to wall themselves off from their surroundings, and in part because they can afford to keep some of their members in a dormant state, untouched by drugs that target growth and function.

The plaque that grows on unbrushed teeth is a biofilm, which explains why antibacterial mouthwashes will never take the place of physically scrubbing it away. When a biofilm forms on an internal implant, surgical removal is often the only choice. The result is tens of thousands of open-heart surgeries and artificial joint removals each year, the latter entailing the traumatic shattering of surrounding bone and of muscle.[14] And these cases represent only the obvious infections, those marked by fever, pain, and malaise. Many if not most implants eventually become home to undercover biofilms. The vast majority of these small bacterial communities never cause problems, for they tend to remain in a dormant state, causing no direct damage and largely evading detection by the immune system. The danger arises when the immune system *does* catch their scent. It then has two options: to tolerate the biofilm's presence, or to allow it to become the kind of maddening thorn that causes chronic inflammation or even septic meltdown.

With the ranks of the bionically enhanced now in the tens of millions, the medical world has responded to the growing danger, though not always wisely. For example, when it became clear that mouth bacteria were frequently taking up residence on implants, many dentists began prescribing a blast of antibiotics before anyone with so much as a bone pin underwent a routine dental cleaning. Follow-up studies showed that this did little to reduce the risk of implant infection and often caused problems such as antibiotic-associated diarrhea and colitis.[15] Still other studies suggest that antibiotics may actually encourage bacteria to form biofilms—as a defensive reaction to the chemical assault.[16]

But what if the implants themselves released antibiotics? Wouldn't that make their surfaces forbidding territory? Some biotech firms are pursuing this controversial goal.[17] Others warn that what seems like a logical approach is a dangerous one. They point to lessons learned from the use of antibiotic beads implanted during surgery to prevent postoperative infections. Years ago, surgeons learned that they had to remove these beads within a few weeks (or use biodegradable versions) or the beads would themselves become encrusted in drug-resistant bacteria.[18] True to warnings, the first antibiotic implant—the Silzone mechanical heart valve—actually increased the risk of heart infection when it was used in patients.[19] Further tests showed that the antimicrobial coating on the device encouraged bacteria to attach.[20]

More sophisticated are recent attempts to discourage biofilm formation by interfering with the "quorum-sensing" signals that members of a bacterial community exchange to coordinate their activities. In the 1990s Princeton biologist Bonnie Bassler showed that bacteria produce signaling molecules that enable them to communicate with a wide variety of species other than their own.[21] Bassler has dubbed this type-2 quorum sensing "bacterial Esperanto." It appears to be essential for the growth of multispecies biofilms, and so presents an inviting target for drugs that would block their formation.

Today Bassler is one of more than a dozen molecular biologists working with companies to translate their new understanding of quorum sensing into biofilm-busting drugs. However, she remains cautious about the wisdom of flooding the body with such chemicals.[22] After all, not all biofilms are bad. Witness the protective biofilm communities formed by lactobacilli in the vagina and by the enormous diversity of bacteria in our guts. In both cases, their disruption clearly predisposes to disease. Recently Bassler and one of her postdoctoral fellows, Karina Xavier, found evidence that some intestinal bacteria may protect against pathogens such as *Vibrio cholerae* by manipulating levels of certain quorum-sensing molecules to "confuse" the enemy.[23] For all these reasons and others unknown, a drug that scrambles quorum sensing throughout the body might have disastrous consequences.

Still, as with conventional antibiotics, it may be possible to target biofilm disrupters to minimize unwanted side effects. For example, orthopedic surgeons report that two kinds of staph—*Staph. aureus* and *Staph. epidermidis*—account for most joint implant infections.[24] The two species use a common quorum-sensing molecule to initiate biofilm formation with members of their own kind—a signal that researchers found they could block with a small protein called RNAIII-inhibiting peptide, or RIP.[25] RIP's discovery gives new hope for "defouling" the many implants that end up coated with staph biofilms.

As for the many other kinds of bacteria that infect rebuilt hips and knees, the most ambitious proposed solution is the creation of a self-diagnosing, self-treating, self-monitoring artificial joint. This futuristic device is the brainchild of bionic engineer and microbiologist Garth Ehrlich, of Drexel University College of Medicine, in Philadelphia.

Bringing together surgeons, microbiologists, and biomechanical engineers, Ehrlich is organizing the development of what he calls the "intelligent" implant, imbued with both antibiotic and biofilm-disrupting drugs that it would deploy in carefully timed bursts. Coordinating this timing would be biosensors able to detect the quorum-sensing signals that microbes exchange when they're "contemplating" communal life.[26] How soon this sophisticated bionic will move from dream to reality will largely depend on advances in our understanding of such "pre-biofilm" signals.

Meanwhile, some bioengineers continue to pursue a potentially simpler solution: the metaphorical "Teflon" implant. Actual Teflon, unfortunately, proves eminently conducive to biofilms. In theory, the ideal implant surface would be too slippery or otherwise physically uncomfortable for microbial loitering. Of the many materials in development, one of the most promising is chitosan, a chemical compound derived from chitin, the sandpapery component of crustacean shells. At Montana State University's Center for Biofilm Engineering, Philip Stewart likens his team's chitosan implant coatings to "a bed of nails." Bacteria that get too close end up "skewered and leaking," he says. "That might not kill them outright, but it certainly discourages them from establishing a foothold."[27] Similarly, biomaterial scientists at the University of Zurich and the University of Texas are collaborating on an implant coating made of polyethylene glycol. Scientists once thought that this molecule presented bacteria with a surface too slippery to grab. The Zurich-Texas team has shown that it physically repels microbes with a microscopic bramble of brushlike fibers. "For them, it's like trying to crawl through a hedge," says Zurich's Jeffrey Hubbell.[28]

FROM SEPSIS TO CHRONIC INFLAMMATION

Deadly sepsis is the immune system's most dangerous response to stubbornly persistent bacteria inside the body, be they in living tissue or adhered to artificial parts. But it's not the only way that the immune system's response to their persistence can spin off in the wrong direction. If the response is chronic low-level inflammation, the result may be one of a wide range of common and debilitating disorders. The most widespread may be atherosclerosis, or hardening of the arteries. Studies suggest that artery-clogging plaque results from an inflammatory response to bacteria lurking in the lining of blood vessels. The enduring mystery is why one person's immune system tolerates their presence while that of another reacts with endless inflammation. The same can be said of *Helicobacter pylori*, the once-ubiquitous stomach bug that causes ulcers in a small fraction of those infected, and of *Chlamydia trachomatis*, the sexually transmitted bug that, in its dormant form, sometimes triggers inflammatory arthritis.

One approach to these so-called stealth infections is to try to eradicate them with antibiotics; another is to fix the human side of the problem. Inflammation-quashing steroids represent the second approach. Unfortunately, they can come with the serious side effect of immune suppression. Some are now pinning their hopes for a second generation of safer anti-inflammatory drugs on experimental treatments that also show promise against sepsis.

"Because severe sepsis and septic shock progress so fast and there are so many things going on that we don't completely understand," says Tracey, "I truly believe that our first successes may come with inflammatory diseases that provide us with more time to intervene." To that end, MedImmune is interested in using Tracey's anti-HMGB1 antibodies in treating patients with rheumatoid arthritis.[29] Tracey is also collaborating in the development of an electrical vagus nerve stimulator that might ease a variety of inflammatory disorders. It might even be possible for people to achieve the same benefit with biofeedback, he says. The idea has precedents. "People can learn to reduce their heart rate by increasing the activity of the vagus nerve," he notes. "It may well be that they can learn to alleviate the symptoms of arthritis, inflammatory bowel disease, and other disorders."

On a more modest level, research shows that simply increasing dietary fats—especially fish oils and olive oil—eases the symptoms of many chronic inflammatory disorders and helps prevent the inflammatory damage seen in atherosclerosis.[30] In 2005 researchers at the University of Maastricht, in the Netherlands, showed why: dietary fat prompts the release of the neurochemical cholecystokinin, which in turn stimulates the vagus nerve.[31] At the least, these findings seem to recommend a "Mediterranean style" diet for those suffering from inflammatory disorders.

One subset of immune-calming treatments deserves its own special category. For while bacteria in the wrong place can trigger damaging inflammation, the same may be true of a relative *lack* of bacteria in the *right* place—that is, in our food, water, and daily environment. This is, after all, the crux of the so-called hygiene hypothesis, and it explains the second widely recognized benefit of probiotic supplements: besides competing with and blocking disease-causing microbes, probiotics appear to ease several kinds of inflammatory disorders.

Among these are inflammatory bowel disorders such as Crohn's disease and ulcerative colitis. Though only partially understood, these disorders appear to result when the lining of the intestinal tract becomes "leaky," allowing bacteria to seep into underlying tissues and spark inflammation. The inflammation exacerbates the seepage, setting up a worsening cycle of painful damage that can turn deadly when ulcers pierce the intestinal wall. What starts the cycle of seepage remains unknown. However, the resulting inflammation can spill over into other parts of the body, such as the joints. This appears to result when antibodies unleashed to fight leaking intestinal bacteria cross-react with healthy cells in joint tissue—mistakenly targeting them for destruction as well.[32]

A variety of probiotics have been shown to provide moderate relief in both inflammatory bowel disease and inflammatory arthritis. The most thoroughly tested of these include Culturelle (*Lactobacillus* GG) and VSL#3, a probiotic mixture developed by gastroenterologists at the University of Bologna. VLS#3 consists of four strains of lactobacilli, three strains of bifidobacteria, and one of the beneficial mouth bug *Streptococcus salivarius*.[33] Both formulas increase the immune system's production of the anti-inflammatory cytokine interleukin-10, reduce gut inflammation, and "tighten up" the normal barriers that keep intestinal bacteria in their place. As mentioned, Belgian researchers recently enhanced this natural gut-calming effect of probiotics by introducing the human gene for interleukin-10 into the cheesemaking bug *Lactococcus lactis*. In 2006 they safety-tested their IL-10-secreting probiotic in ten patients with Crohn's disease, making it the first transgenic microbe used in humans. Though the weeklong study was too small and brief to prove effectiveness, the ten patients reported an easing in their usual symptoms of diarrhea, bloating, and arthritis pain.[34]

Unaltered probiotics such as *Lactobacillus* GG, with its anti-inflammatory powers, may also help prevent the development of eczema and other allergic disorders. In a recent clinical trial researchers gave the supplement to pregnant women from allergy-prone families, while a comparison group received a placebo. Their babies likewise received the supplement (mixed in breast milk or formula). At six months, half as many of the infants who received the probiotic had developed eczema as had those in the comparison group. A follow-up study showed that the protection was still in effect when these children reached four years of age.[35] In still other tests *Lactobacillus GG* helped relieve eczema in babies who had already developed the condition on the heels of food allergies.[36]

In animal studies, *Lactobacillus GG*, VSL#3, and other probiotics have also proven effective in preventing autoimmune diseases such as diabetes.[37] In 2006 Swedish researchers at Linkoping University became the first to advance such a treatment to human trials, enrolling two hundred newborns with genetic markers showing them at high risk of autoimmune diabetes. The supplement in their test consists of four live organisms: *Lactobacillus GG*, the closely related *Lactobacillus rhamnosus* LC705, a bifidobacterium, and a strain of the Swiss-cheese culture bacterium *Propionibacterium freudenreichii*.[38] When the last of the children reach their fifth birthdays—in June 2010—the researchers will break the code that reveals which received the probiotic and which received a placebo, and assess whether the probiotics were of benefit.

Considering the immune system's rapid development during infancy, giving beneficial bacteria to infants may well have dramatic benefits. Less clear is whether probiotics can benefit older children and adults, especially those who have already developed an inflammatory disorder. Hygiene hypothesis researchers such as Harvard's Umetsu and University College London's Rook and Stanford pioneered efforts to gain a bigger immunological punch from beneficial bacteria. They selected species with unusually potent immunological effects and then delivered killed cells as vaccines (listeria in Umetsu's vaccine and mycobacteria in Rook and Stanford's). In 2006 two other research teams leapfrogged over the groundbreaking work of these three pioneers. The result: two sets of bioengineered vaccines that borrow from bacteria but don't actually contain them.

IMMUNOBUG IMMUNODRUGS

On a fall afternoon in 2006, Swiss immunologist Martin Bachmann is in an expansive mood as he steers his silver Audi convertible east along the high-speed motorway connecting Zurich and Bern. Cytos Biotechnology, the company that Bachmann helped build, has just released its third-quarter report to investors. The report is a particularly good one, thanks to the success of two pilot clinical trials with the "Immunodrug" that Bachmann began developing as a doctoral student at the University of Zurich.

"I quite enjoy it," he says of his jump from academic to corporate-based research. "I went from having one technician in a little lab to a big group, very organized, very serious, and still good basic research." As for the successful pilot studies, one involved ten longtime hay fever sufferers who received six weekly injections of Bachmann's Immunodrug. They remained symptom-free for the six-month duration of the trial. "We saw a hundredfold increase in tolerance," he exults. That is, when a nurse subjected the immunized volunteers to a "nasal provocation test" (essentially repeated snootfuls of pollen), they withstood a hundred times more allergen before getting snotty than they had before receiving the treatment.[39] The company's more modest hope had been to produce a tenfold increase in tolerance. "This is super significant," Bachmann exults with a rap on the steering wheel. "On a rainy day in summer, you have around ten times less pollen in the air than on a sunny day, and that's enough to make most people's hay fever disappear. To produce a hundredfold increase in tolerance is, for all practical purposes, a cure."

The year's other encouraging study involved twenty dust-mite-allergic asthmatics and produced similar benefits, with seventeen remaining free of allergy symptoms, even when given eyedrops filled with concentrated dust-mite allergen. Even better, says Bachmann, nineteen out of the twenty experienced relief from their asthma.[40]

How does Bachmann's Immunodrug work? "As a doctoral student I began looking at why viruses produced such potent antibody responses," he says of its inception. He found the key trait to be the highly repetitive, almost crystalline structure of a virus's protein capsule.[41] The immune system immediately recognizes this pattern as foreign, as there's nothing remotely like it in the human body. Next Bachmann found he could manufacture his own viruslike particles using viral proteins. "I immediately thought that this would be a great way to design a vaccine that produced a very good antibody response to whatever antigen one attached to it."

But while Bachmann's viruslike particles grabbed the attention of one arm of the immune system—the B cells that produce antibodies—it failed to make a lasting impression on the T cells needed to produce lasting immunity. That was when Bachmann turned to the bacterial kingdom for the immune response he needed. He borrowed a segment of DNA from mycobacteria, long known to induce a strong immune-calming response. In another bit of biochemical sleight of hand, Bachmann slipped the mycobacterial DNA *inside* his viruslike particles, in much the same way that an actual virus would package its own DNA.[42]

"What's elegant," he says, "is how packaging the DNA inside the viruslike particles protects it from being degraded and delivers it to the right immune cells." Specifically, the pseudo-virus delivers the bacterial DNA to dendritic cells, which play a crucial role in influencing the response of the body's T cells. The resulting virus-bacteria chimera became Cytos Biotechnology's patented "Immunodrug Platform." Intriguingly, in the dust-mite allergy trial, Bachmann mixed his viral-bacterial particles with the offending allergen, with the intent of provoking an allergen-specific response. In the hay fever trial he included no allergen but produced an equally beneficial effect. "I don't think it matters," he says. "It may be that the mycobacteria by themselves are potently anti-allergenic and what we've done is mimic a naturally occurring mycobacterial infection." Still, for all Bachmann's excitement over early results, his therapies must still prove themselves in much larger clinical trials. If they do, he may have a marketable product as early as 2010.

TWEAKING THE BUG

Mycobacterial DNA likewise figures prominently in the immune-modulating vaccines being developed by Dynavax, a Berkeley, California, biotech firm launched by immunologist Eyal Raz. Unlike

Bachmann, Raz has chosen to remain in academia, at the University of California San Diego, after assembling a scientific team to translate his research into medical products at Dynavax.

Raz's early research, in the mid-1990s, helped establish what it was about mycobacterial DNA that so strongly grabs the immune system's attention. It turned out to be the sequence and relative abundance of two out of the four nucleotide "letters" (cytosine and guanine) that the bacterium uses to spell out its genes.[43] Since then Raz has been building his own cytosine-guanine, or CpG, sequences, all of them based on mycobacterial DNA, with subtle variations. These varied sequences appear to produce profoundly different responses from the immune system. Some turn out to be strongly anti-allergenic; others seem to ameliorate autoimmune reactions, inflammatory bowel disease, even deadly sepsis — at least in lab animals.[44]

Treatments for humans took a leap forward in 2006, when *The New England Journal of Medicine* published the results of Dynavax's first patient trial using a CpG vaccine for hay fever.[45] Fourteen of the study's twenty-five volunteers received six weekly injections of the active ingredient: a sequence of chemically altered mycobacteria DNA chemically linked to the allergen in ragweed pollen. The rest of the volunteers got dummy shots. The treatment halved the volunteers' symptoms over the following *two* ragweed seasons, enough that none of them had to reach for the antihistamines and decongestants that they previously needed to get through each fall.

INTO THE FUTURE

On a continuing basis, our immune systems engage with the bacterial world around us. For the most part, the interaction proves beneficial, keeping microbes in their place and maintaining the immune system in a state of calm alertness. As scientists deepen our understanding of this dynamic give-and-take, we gain ever more opportunities to intervene when something goes awry. With rates of both deadly sepsis and chronic inflammatory disease on the rise, the need has never been greater.

EMBRACING THE MICROBIOME

Since the dawn of civilization, the demon of pestilence has been a part of our lives and fears. Sanitation and antibiotics gave us our first powerful weapons against this great foe. But we have wielded them crudely, without appreciation either for the role that bacteria play in maintaining our health or for their infinite capacity to adapt to whatever poisons we throw at them. Though glimmers of understanding go back as far as Pasteur—and his conviction that life without germs would be impossible—efforts to distinguish the "good guys" from the bad were largely lost in the exultation of microbe hunting and the apparent conquest of one disease-causing foe after another.

The rapid rise of drug-resistant superbugs has been our rude awakening. "From an evolutionary point of view, the bacteria have always had the advantage," says Joshua Lederberg, the Nobel laureate who with his former wife, Esther, unfolded the mechanisms of what he now calls the "world wide web" of microbial gene exchange. "Bacteria can multiply and evolve a million times more rapidly than we can," he notes.[1] They don't quibble about who belongs to what species when it comes to swapping the genes they need to thwart our antibiotics and, if it's in their interest, to wipe us out.

So why haven't they?

"They need us, just as surely as we need them," says Lederberg. "The bug that kills its host is at a dead end." Admittedly, we continue to encourage the rise of such rogue microbes by giving them new ways to jump to new hosts faster than they waste the old ones. Thanks to the quirks of modern food distribution, for example, a deadly microbe like *E. coli* 0157:H7 can now spread from a single contaminated farm field to dinner plates across an entire continent in a day.

More than ever, perhaps, we need to keep the living armor of our microflora *firmly* in place. Ironically, a new popular awareness of these, our most intimate associates, has come from headlines associating gut bacteria with obesity. Hopefully, people don't come away from such media reports with the idea that fatness is something to cure with antibiotics, but instead gain an appreciation that our gut bacteria have always functioned as a vital digestive organ (with some species being more efficient at calorie extraction than others).

"It would broaden our horizons if we started thinking of a human as more than a single organism," Lederberg elaborates. "It is a super-organism that includes much more than our human cells." Lederberg calls this cohabitation of human and microbial cells the "microbiome" as he urges a new generation of microbiologists to further our understanding of how it unites two vastly different kingdoms of life into an integrated whole. "I've never said we should never kill a bug," Lederberg quickly adds. "After all, the bug doesn't have a dictum to never kill a person, even if it's sealing its own doom. What's important is that we're better off aspiring to a relationship of symbiotic coexistence."

In this context, the treatment of infectious disease becomes less a war on an invisible enemy than a restoration of balance—sometimes fixing the host, as when the immune system becomes overly aggressive or neglectful of its duty to keep each microbe in its place; other times bolstering our microflora's ability to perform its many vital functions— digestion, the rebuffing of pathogens, and the taming of inflammation being the most obvious. Are we ready for such a revolutionary new way of viewing human biology? It may well hold the key to our continued health and survival in what always has been, and always will be, a bacterial world.

NOTES

PROLOGUE: A GOOD WAR GONE BAD

1. Last name omitted, at the request of the family, to protect Daniel's privacy.

PART I: THE WAR ON GERMS

Part opener: Paul De Kruif, *The Microbe Hunters* (Harcourt, Brace & World, 1926), 354.

1. Girolamo Fracastoro Veronae, *Syphilis sive morbus gallicus* (Stefano dei Nicolini de Sabbio e Fratelli, 1530).
2. Nahum Tate's 1686 English translation is *Syphilis, or a poetical history of the French disease* (Early English Books, 1641–1700), 1229:22.
3. *Plasmodium falciparum*, as a protozoan microbe, is more closely related to plants and animals than to bacteria.
4. The complete epic can be found online at www.ancienttexts.org/library/mesopotamian/gilgamesh/. For reference to Ura, god of pestilence, see Tablet XI.
5. *De sympathia et antipathia rerum. De contagione et contagiosis morbis et curatione libri tres* (Venice: Apud heredes Lucaentonii Juntae Florentini, 1546).
6. Thomas Parran, *Shadow on the Land: Syphilis* (New York: Reynal & Hitchcock, 1937), 46–47, 168, a pre–antibiotic era treatise on diagnosis and treatment. See also the Cambridge and Cantabria historians Jon Arrizabalaga, John Henderson, and Roger French, *The Great Pox: The French Disease in Renaissance Europe* (New Haven, Conn.: Yale University Press, 1997), 142–44 (on mercury) and 244–51 (on Fracastoro and his theories of contagion).
7. Some believe this practice to be the origin of the children's rhyme "Ring around the rosy [bubonic plague's red rash]; a pocketful of posies; ashes, ashes, we all fall down [dead]."
8. Diary of Samuel Pepys, excerpts online at www.pepys.info/1665/1665.html.
9. Robert Hooke, *Micrographia: or, Some physiological descriptions of minute bodies made by magnifying glasses* (London, 1665), searchable full text available online through Project Gutenberg, at www.gutenberg.org/files/15491/15491-h/15491-h.htm.
10. Some historians argue that Leeuwenhoek saw *Micrographia* during a visit to London in 1668. For the most part, I have based my account on the deeply researched biography written by English microbiologist Clifford Dobell, arguably the most authoritative and scientifically insightful of the many Leeuwenhoek translations and biographies: Clifford Dobell, *Antony van Leeuwenhoek and His "Little Animals," Collected, Translated and Edited from His Printed Works, Unpublished Manuscripts and Contemporary Records* (New York: Dover, 1932).
11. The "eukaryotic" cells of microscopic algae, protozoa, and fungi, like those of all

plants and animals, are much larger and more complex than the "prokaryotic" cells of bacteria.

12. The twentieth-century bacteriologist Theodor Rosebury identified "the biggest sort" as spirochete bacteria rotating so fast that Leeuwenhoek could not discern their undulating coils. "The second sort" were most likely spirillum bacteria; their teardrop shape and wild activity are familiar to high school biology students who repeat Leeuwenhoek's classic experiment with modern microscopes. As for "the third sort": taking into account that Leeuwenhoek could barely make out these last bacteria, Rosebury identified them as an assortment of streptococci, staphylococci, micrococci, and rod-shaped bacilli—all abundant in the human mouth. Rosebury identified those in Figure F as a variety of filamentous mouth bacteria, including *Actinomyces* and *Leptotrichia*, and those in Figure G as a spirochete or a variant of the tumblers seen in Figures B through D. Dobell, *Leuwenhoek*, 237–44.

13. Dobell, *Leuwenhoek*, 243.

14. Alexander Gordon, A *Treatise on the Epidemic Puerperal Fever of Aberdeen* (London: G. G. and J. Robinson, 1795); excerpted in Peter Dunn, "Perinatal Lessons from the Past: Dr. Alexander Gordon and Contagious Puerperal Fever," *Archives of Diseases of Children and Fetal Neonatal Education* 78 (1998), F232–33.

15. Oliver Wendell Holmes, "Contagiousness of Puerperal Fever," *New England Quarterly Journal of Medicine* (1843). The full text is available in *Scientific Papers; Physiology, Medicine, Surgery, Geology, with Introductions, Notes and Illustrations* (New York: P. F. Collier & Son, c1910); *The Harvard Classics* v. 38; and online at biotech.law.lsu.edu/cphl/history/articles/pf_holmes.htm.

16. Francesco Redi disproved the theory of the spontaneous generation of insects in 1667, with his classic experiment using screens to keep flies away from rotting meat. But Redi and his followers still clung to the belief that intestinal worms and smaller parasites arose in this way.

17. For many decades after it was coined, the word *bacteria* continued to be used interchangeably with over a dozen general terms such as *animalcules, infusoria, mucors, vibrios, monads, viruses,* and *fungi,* without any of the specificity that some of these terms have today.

18. Benjamin Marten, A *New Theory of Consumptions: More Especially of a Phthisis, or Consumption of the Lungs* (London, 1720), 40–41.

19. *The Devil Upon Two Sticks,* in *The Dramatic Works of Samuel Foote, Esq.* (New York: Benjamin Blom, 1968) 2:50–51.

20. Friedrich Gustav Jacob Henle, *Über Miasmen und Contagien und von miasmatisch-contagiosen Krankheiten*; English translation in *Bulletin of the History of Medicine* 6 (1936), 911–83.

21. A eukaryotic cell has a nucleus, or membrane-bound envelope, containing the genetic structures we call chromosomes (tightly coiled strands of deoxyribonucleic acid, or DNA) and also contains a number of other membrane-enclosed organelles such as mitochondria and, in photosynthetic organisms, chloroplasts. The simpler, prokaryotic cell of a bacterium lacks a nucleus and organelles and typically contains a single chromosome that floats freely in the cell's cytoplasm, or cell fluid.

22. "Together with the more important lessons, which you endeavour to teach all the poor whom you visit, it would be a deed of charity to teach them two things more, which they are generally little acquainted with—industry and cleanliness. It was said by a pious man, 'Cleanliness is next to godliness.'" John Wesley, Sermon 98,

"On Visiting the Sick," May 23, 1786 (Christian Classics Ethereal Library, www.ccel.org).

23. Florence Nightingale is quoted in Suellen Hoy's delightful *Chasing Dirt: The American Pursuit of Cleanliness* (Oxford: Oxford University Press, 1995).

24. Frederick S. Odell, "The Sewerage of Memphis with Discussions," *Transactions of the American Society of Civil Engineers* 9 (February 1881), 24–26.

25. George F. Waring, Jr., *Street Cleaning and the Disposal of a City's Wastes* (New York: Doubleday & McClure, 1897), 13–14.

26. Jacob Riis, *The Battle with the Slum* (1902), chap. 11; full text online at Bartleby.com, www.aol.bartleby.com/175/11.html.

27. Waring, *Street Cleaning*, 20–21.

28. These U.S. life expectancy figures are from the Department of Health and Human Services, National Center for Health Statistics, *National Vital Statistics Reports* 52, no. 3, September 18, 2003. The U.K. mortality figures are from www.mortality.org.

29. Louis Pasteur, "Observations relatives à la note precedente de M. Duclaux," *Comptes rendus de l'Academie des Sciences* 100 (1885), 68. Pasteur was commenting on his student Émile Duclaux's experiment comparing plants growing in sterile soil versus soil inoculated with beneficial, nitrogen-fixing bacteria. For a deeply researched and informative analysis in English, see Canadian science historian Jan Sapp's *Evolution by Association* (Oxford: Oxford University Press, 1994), chaps. 6 and 7.

30. Elie Metchnikoff, *The Nature of Man* (New York: G. P. Putnam's Sons, 1903), 252–57, and *The Prolongation of Life* (New York: G. P. Putnam's Sons, 1908), 61–83. Ironically, later generations would dub Metchnikoff the "father of probiotics" (probiotics being the study and use of beneficial bacteria) because one of his favorite methods for battling the "chronic poisoning from an abundant intestinal flora" was to drink liters of sour milk cultured with lactic-acid bacteria. Their acid, he believed, was "inimical to the growth of the bacteria of putrefaction."

31. Thomas D. Luckey, *Germfree Life and Gnotobiology* (New York: CRC Press, 1963).

32. Macfarlane Burnet, "Preface to the Third Edition," *Natural History of Infectious Disease* (Cambridge: Cambridge University Press, 1962).

PART II: LIFE ON MAN

Part opener: Antoni van Leeuwenhoek, Letter to the Royal Society of London, 1683, reprinted in Clifford Dobell, *Antony van Leeuwenhoek and His "Little Animals"; Being Some Account of the Father of Protozoology and Bacteriology and His Multifarious Discoveries in These Disciplines* (Mineola, N.Y.: Dover, 1960).

1. Theodor Rosebury, *Microorganisms Indigenous to Man* (New York: McGraw-Hill, 1962).

2. Theodor Rosebury, *Life on Man* (New York: Viking Press, 1969), 8.

3. Ibid., xv.

4. The threat of neisseria- and chlamydia-related blindness is the reason hospitals drip antibiotics into the eyes of newborns. Untreated eye infections can cause blindness over the course of several weeks.

5. Yu-Li Song et al., "Identification of and Hydrogen Peroxide Production by Fecal and Vaginal Lactobacilli Isolated from Japanese Women and Newborn Infants," *Journal of Clinical Microbiology* 37 (1999), 3062–64.

6. M. Bayo et al., "Vaginal Microbiota in Healthy Pregnant Women and Prenatal Screening of Group B Streptococci (BGS)," *International Microbiology* 5 (2002), 87–90.

7. This information on the colonization of the human mouth is from Rosebury, *Microbes Indigenous*, 327–31, as confirmed and elaborated in Michael Wilson, *Microbial Inhabitants of Humans: Their Ecology and Role in Health and Disease* (New York: Cambridge University Press, 2004), 318–60.

8. *N. meningitidis* infects around one in a thousand infants; its mortality rate has dropped from over 75 percent in the 1980s to less than 3 percent today, thanks to improvements in early recognition and intensive care treatment.

9. R. J. Berkowitz et al., "Maternal Salivary Levels of *Streptococcus mutans* and Primary Oral Infection of Infants," *Archives of Oral Biology* 26 (1981), 147–49.

10. Mary J. Marples, *The Ecology of the Human Skin* (Springfield, Ill.: Charles C. Thomas, 1965); see also Mary Marples, "Life on the Human Skin," *Scientific American* 220 (January 1969), 108–15.

11. Much of the definitive research on the chemical makeup and pheromonelike effects of armpit "extracts" comes out of the laboratory of George Preti, at the Monell Chemical Senses Center, an independent scientific research institute in Philadelphia. For a good review of the work, see Charles Wysocki and George Preti, "Facts, Fallacies, Fears, and Frustrations with Human Pheromones," *Anatomical Record* 281A (2004), 1201–11.

12. Marples, "Life on Human Skin."

13. Wystan Hugh Auden, "A New Year Greeting," read at the Poetry International Festival, 1969, published in *Scientific American* 220 (December 1969), 130.

14. Thomas Luckey, "Discussion: Intestinal Flora," NASA Conference on Nutrition in Space and Related Waste Problems, University of South Florida, April 27–30, 1964, NASA document number SP-70; Thomas Luckey, interviews by author, July 2006; and T. D. Luckey, "Potential Microbic Shock in Manned Aerospace Systems," *Aerospace Medicine* 37 (1966), 1223–28.

15. Luckey, "Studies on Germfree Animals," *Journal of the Chiba Medical Society* 35 (1959), 1–24, as reported in "Potential Microbic Shock."

16. Lorraine Gall and Phyllis Riely, "Effect of Diet and Atmosphere on Intestinal and Skin Flora," in *A Report of the Physiological, Psychological, and Bacteriological Aspects of 20 Days in Full Pressure Suits, 20 Days at 27,000 Feet on 100 Percent Oxygen and 34 Days of Confinement*, NASA Technical Report NAEC-ACELL-535, published April 1, 1966.

17. V. M. Shilov et al., "Changes in the Microflora of Man During Long-term Confinement," *Life Sciences and Space Research* 9 (1971), 43–49.

18. *Skylab Medical Experiments Altitude Test (SMEAT)* (1973), NASA Technical Report TM X-58115, Johnson Space Center, Houston, Tex., and NASA contract 9-12601; Charles Berry, "Proposal for Ground-Based Manned Chamber Study for the Skylab Program," March 24, 1970, as quoted in *A History of Living and Working in Space: A History of Skylab*, NASA Historical Report SP-4208, chap. 8.

19. Moore had been improving on the anaerobic culture techniques developed by pioneering rumen biologist Robert Hungate, at the University of California at Davis.

In essence, the Hungate Method involved growing anaerobic bacteria in a thin film of specially prepared agar on the inner surface of a sealed tube from which all oxygen had been purged with a flush of carbon dioxide.

20. Peg Holdeman Moore, interviews by the author, August–September 2006.

21. Robert Smibert, interview by John Hess, Virginia Tech Oral History Project, June 19, 2002.

22. Moore, interview by the author; Peg Holdeman, "Human Fecal Flora: Variation in Bacterial Composition with Individuals and a Possible Effect of Emotional Stress," *Applied and Environmental Microbiology* 31 (1976), 359–75.

23. L. V. Holdeman and W.E.C. Moore, "A New Genus, *Coprococcus*, Twelve New Species, and Amended Descriptions of Four Previously Described Species of Bacteria from Human Feces," *International Journal of Systematic Bacteriology* 24 (1974), 260–77.

24. Thanks to California midwife and author Ronnie Falcao (www.gentlebirth.org) for her descriptions of this aspect of the birthing process.

25. Thierry Wirth et al., "Distinguishing Human Ethnic Groups by Means of Sequences from *H. pylori*: Lessons from Ladakh," *Proceedings of the National Academy of Sciences* 101 (2004), 4746–51; Shan-Rui Han et al., "*Helicobacter pylori*: Clonal Population Structure and Restricted Transmission Within Families Revealed by Molecular Typing," *Journal of Clinical Microbiology* 38 (2000), 3646–51.

26. Denys Jennings, "Perforated Peptic Ulcer: Changes in Age-Incidence and Sex-Distribution in the Last 150 Years," *Lancet* 235 (1940), 395–98.

27. W. O. Huston, "The American Disease," *Columbus Medical Journal* 16 (1896), 1–7.

28. B. Marshall and R. Warren, "Unidentified Curved Bacilli in the Stomach of Patients with Gastritis and Peptic Ulceration," *Lancet* 8390 (1984), 1311–15.

29. S. A. Dowsett et al., "*Helicobacter pylori* Infection in Indigenous Families of Central America: Serostatus and Oral and Fingernail Carriage," *Journal of Clinical Microbiology* 37 (1999), 2456–60.

30. Martin Blaser, "In a World of Black and White, *Helicobacter pylori* Is Gray," *Annals of Internal Medicine* 130 (1999), 695–97.

31. The upper surface of the so-called M cells that overlie a Peyer's patch are riddled with pockets that can each comfortably accommodate several bacteria. Once bacteria enter a pocket, it closes and migrates to deliver its passengers to a larger crypt on the cell's opposite, or internal, surface, where it meets a horde of waiting immune cells. As for the "M" in the name M cells, I have yet to definitively track down its meaning. Perhaps it refers to "mesenteric," a broad reference to all things intestinal? I'd be grateful for further information (livingwithmicrobes @jessicasachs.com).

32. Thomas MacDonald and Giovanni Monteleone, "Immunity, Inflammation, and Allergy in the Gut," *Science* 307 (March 2005), 1920–25; Ralph Steinman et al., "Tolerogenic Dendritic Cells," *Annual Reviews in Immunology* 21 (2003), 685–711; Daniel Hawiger et al., "Dendritic Cells Induce Peripheral T Cell Unresponsiveness Under Steady State Conditions In Vivo," *Journal of Experimental Medicine* 194 (September 2001), 769–79; Ralph Steinman et al., "Dendritic Cell Function In Vivo During the Steady State: A Role in Peripheral Tolerance," *Annals of the New York Academy of Sciences* 987 (2003), 15–25.

33. Andrew Macpherson and Therese Uhr, "Induction of Protective IgA by Intestinal Dendritic Cells Carrying Commensal Bacteria," *Science* 303 (March 2004), 1662–65.

34. Helena Tlaskalova-Hogenova et al., "Commensal Bacteria (Normal Microflora), Mucosal Immunity and Chronic Inflammatory and Autoimmune Diseases," *Immunology Letters* 93 (2004), 97–108.

35. Henri Tissier, "Repartition des microbes dans l'intestin du nourisson," *Annals of the Pasteur Institute* 19 (1905), 109–23, as cited in G. W. Tannock, "The Acquisition of the Normal Microflora of the Gastrointestinal Tract," in S.A.W. Gibson, ed., *Human Health: The Contribution of Microorganisms* (New York: Springer-Verlag, 1994), 1–16.

36. American Academy of Pediatrics, "Human Milk," *Red Book* (January 2003), 117–23.

37. WHO Collaborative Study Team, "Effect of Breastfeeding on Infant and Child Mortality Due to Infectious Diseases in Less-Developed Countries: A Pooled Analysis," *Lancet* 355 (2000), 451–55.

38. Aimen Chen and Walter Rogan, "Breastfeeding and the Risk of Postneonatal Death in the United States," *Pediatrics* 113 (2004), 435–39.

39. Our understanding of these processes comes in large part from comparisons of normal (bacteria-colonized) and germ-free lab animals, the latter artificially delivered and painstakingly raised in sterilized environments.

40. Wilson, *Microbial Inhabitants of Humans*, 264–81; Abigail Salyers and Dixie Whitt, *Microbiology* (Bethesda, Md.: Fitzgerald Science Press, 2001), 225–28.

41. "Recent Trends in Mortality Rates for Four Major Cancers, by Sex and Race/Ethnicity, United States, 1990–1998," *Morbidity and Mortality Weekly Report* 51 (2002), 49–53. Curiously, colon cancer rates remain low in regions of the world that lack modern water sanitation.

42. Abigail Salyers and Mark Pajeau, "Competitiveness of Different Polysaccharide Utilization Mutants of *Bacteroides thetaiotaomicron* in the Intestinal Tracts of Germfree Mice," *Applied and Environmental Microbiology* 55 (1989), 2572–78.

43. Lynn Bry et al., "A Model of Host-Microbial Interactions in an Open Mammalian Ecosystem," *Science* 273 (1996), 1380–83; and Lora Hooper et al., "A Molecular Sensor That Allows a Gut Commensal to Control Its Nutrient Foundation in a Competitive Ecosystem," *Proceedings of the National Academy of Sciences* 96 (1999), 9833–38.

44. The DNA sequence of an active, or expressing, gene gets transcribed into complementary molecules called messenger RNA, which in turn become templates for the cell to assemble the enzymes and other proteins that perform all its functions.

45. Lora Hooper et al., "Molecular Analysis of Commensal Host-Microbial Relationships in the Intestine," *Science* 291 (2001), 881–84.

46. Thaddeus Stappenbeck et al., "Developmental Regulation of Intestinal Angiogenesis by Indigenous Microbes via Paneth Cells," *Proceedings of the National Academy of Sciences* 99 (2003), 15451–55.

47. Jian Xu et al., "A Genomic View of the Human-*Bacteroides thetaiotaomicron* Symbiosis," *Science* 299 (2003), 2074–76.

48. Fredrik Bäckhed et al., "The Gut Microbiota as an Environmental Factor That Regulates Fat Storage," *Proceedings of the National Academy of Sciences* 101 (2004), 15718–23.

49. Justin Sonnenburg et al., "Getting a Grip on Things: How Do Communities of Bacterial Symbionts Become Established in Our Intestine?" *Nature Immunology* 5 (2004), 569–73.

50. Ruth Ley et al., "Human Gut Microbes Associated with Obesity," *Nature* 444 (2006), 1022–23; Peter Turnbaugh et al., "An Obesity-Associated Gut Microbiome with Increased Capacity for Energy Harvest," *Nature* 444 (2006), 1027–31.

51. Buck Samuel and Jeffrey Gordon, "A Humanized Gnotobiotic Mouse Model of Host–Archaeal–Bacterial Mutualism," *Proceedings of the National Academy of Sciences* 103 (2006), 10011–16.

52. Carl Woese and G. E. Fox, "Phylogenetic Structure of the Prokaryotic Domain: The Primary Kingdoms," *Proceedings of the National Academy of Sciences* 74 (1977), 5088–90.

53. Norman Pace, "The Analysis of Natural Microbial Populations by Ribosomal RNA Sequences," *Advances in Microbial Ecology* 9 (1986), 1–55.

54. Li Weng et al., "Application of Sequence-Based Methods in Human Microbial Ecology," *Genome Research* 16 (2006), 316–22.

55. P. Hugenholts, B. M. Goebel, and N. R. Pace, "Impact of Culture-Independent Studies on the Emerging Phylogenetic View of Bacterial Diversity," *Journal of Bacteriology* 180 (1998), 4765–74.

56. Pace, "Analysis of Natural Microbial Populations."

57. Ken Wilson et al., "Phylogeny of the Whipple's Disease–Associated Bacterium," *Lancet* 338 (1991), 474–75; David Relman et al., "Identification of the Uncultured Bacillus of Whipple's Disease," *New England Journal of Medicine* 327 (1992), 293–301.

58. Ian Kroes, Paul Lepp, and David Relman, "Bacterial Diversity Within the Human Subgingival Crevice," *Proceedings of the National Academy of Sciences* 96 (1999), 14547–52.

59. Bruce Paster et al., "Bacterial Diversity in Human Subgingival Plaque," *Journal of Bacteriology* 183 (2001), 3770–83; C. E. Kazor et al., "Diversity of Bacterial Populations on the Tongue Dorsa of Patients with Halitosis and Healthy Patients," *Journal of Clinical Microbiology* 41 (2003), 558–63; Bruce Paster et al., "The Breadth of Bacterial Diversity in the Human Periodontal Pocket and Other Oral Sites," *Periodontology 2000* 42 (2006), 80–87.

60. Steven Gill et al., "Metagenomic Analysis of the Human Distal Gut Microbiome," *Science* 312 (2006), 1355–59.

61. David Relman, interviews by the author, February–March 2005, July–August 2006.

62. Alan Hudson, interviews by the author, January–March 2005.

63. H. C. Gerard et al., "Chromosomal DNA from a Variety of Bacterial Species Is Present in Synovial Tissue from Patients with Various Forms of Arthritis," *Arthritis and Rheumatology* 44 (2001), 1689–97.

64. B. J. Balin et al., "Identification and Localization of *Chlamydia pneumoniae* in the Alzheimer's Brain," *Medical Microbiology and Immunology* 187 (1998), 23–42.

65. John Grayston, "Does *Chlamydia pneumoniae* Cause Atherosclerosis?" *Archives of Surgery* 134 (1999), 930–34; E. V. Kozarov et al., "Human Atherosclerotic Plaque Contains Viable Invasive *Actinobacillus actinomycetemcomitans* and *Porphyromonas gingivalis*," *Atherosclerosis, Thrombosis and Vascular Biology* 25 (2005), 17–18.

66. Christopher Cannon et al., "Antibiotic Treatment of *Chlamydia pneumoniae* After Acute Coronary Syndrome," *New England Journal of Medicine* 352 (2005), 1646–54.
67. ActivBiotics. Company profile at www.activbiotics.com/companyProfile/index .html.
68. J. R. O'Dell et al., "Treatment of Early Rheumatoid Arthritis with Minocycline or Placebo: Results of a Randomized, Double-Blind, Placebo-Controlled Trial," *Arthritis and Rheumatism* 40 (1997), 842–48.
69. Susan Swedo et al., "Pediatric Autoimmune Neuropsychiatric Disorders Associated with Streptococcal Infections: Clinical Description of the First 50 Cases," *American Journal of Psychiatry* 155 (1998), 264–71.

PART III: TOO CLEAN?
Part opener: Charles Dudley Warner, *My Summer in a Garden* (1870; reprinted by Modern Library, 2002), 4.
1. Joseph Kirsner, "Historical Origins of Current IBD Concepts," *World Journal of Gastroenterology* 7 (2001), 175–84; Richard Logan, "Inflammatory Bowel Disease Incidence: Up, Down or Unchanged?" *Gut* 42 (1998), 309–11.
2. I first heard the term "citizenship disease" from a first-generation, Nigerian-American neighbor, who says he has heard Russian immigrants use the term as well.
3. Hasan Arshad, Suresh Babu, and Stephen Holdate, "History of Allergy," in Martin Dunitz, ed., *Anti-IgE Therapy for Asthma and Allergy* (London: Taylor & Francis, 2001).
4. M. B. Emanuel, "Hay Fever, a Post-industrial Revolution Epidemic: A History of Its Growth During the 19th Century," *Clinical Allergy* 18 (1988), 295–304.
5. William Heberden, *Commentarii de moroborum historia et curatione* (London: 1802), chap. 11.
6. John Bostock, "On the *Catarrhus aestivus* or Summer Catarrh," *Medico-Chirurgical Transactions* 14 (1828), 437–46.
7. Charles Blackley, *Experimental Researches on the Causes and Nature of* Cattarrhus aestivus *(Hay-Fever or Hay-Asthma)* (London: Balliere, Tindeall and Cox, 1873).
8. Emanuel, "Hay Fever."
9. Alfred Neugut et al., "Anaphylaxis in the United States," *Archives of Internal Medicine* 161 (2001), 15–21; Hugh Sampson, "Food Allergy. Part 1: Immunopathogenesis and Clinical Disorders," *Journal of Allergy and Clinical Immunology* 103 (1999), 717–28.
10. Spyros Marketos and Constantine Ballas, "Bronchial Asthma in the Medical Literature of Greek Antiquity," *Journal of Asthma* 19 (1982), 263–69.
11. Oystein Ore, *Cardano the Gambling Scholar* (Princeton: Princeton University Press, 1953).
12. Asthma and Allergy Foundation of America, *Chronic Conditions: A Challenge for the 21st Century* (National Academy on an Aging Society, 2000).
13. B. Taylor et al., "Changes in the Reported Prevalence of Childhood Eczema Since the 1939–45 War," *Lancet* 2 (1984), 1255–57.
14. D. P. Strachan and R. A. Elton, "Relationship Between Respiratory Morbidity in Children and the Home Environment," *Family Practice* 3 (1986), 137–42.

15. David Strachan, "Hay Fever, Hygiene, and Household Size," *British Medical Journal* 299 (1989), 1259–60.
16. C. Svanes et al., "Childhood Environment and Adult Atopy: Results from the European Community Respiratory Health Survey," *Journal of Allergy and Clinical Immunology* 103 (1999), 415–20.
17. David Strachan, "Family Size, Infection and Atopy," *Thorax* 55 (2000), S2–S10.
18. S. L. Prescott et al., "Development of Allergen-Specific T-cell Memory in Atopic and Normal Children," *Lancet* 353 (1999), 196–200; F. D. Martinez, "Maturation of Immune Responses at the Beginning of Asthma," *Journal of Allergy and Clinical Immunology* 103 (1999), 355–61.
19. "Plagued by Cures," *Economist*, November 22, 1997, 95.
20. Thomas Ball et al., "Siblings, Day-Care Attendance and the Risk of Asthma and Wheezing During Childhood," *New England Journal of Medicine* 343 (2000), 538–43.
21. David Strachan et al., "Family Structure, Neonatal Infection and Hay Fever in Adolescence," *Archives of Diseases in Childhood* 74 (1996), 422–26.
22. K. Wickens et al., "Antibiotic Use in Early Childhood and the Development of Asthma," *Clinical and Experimental Allergy* 6 (1999), 766–71; J. H. Droste et al., "Does the Use of Antibiotics in Early Childhood Increase the Risk of Asthma and Allergic Disease?" *Clinical and Experimental Allergy* 11 (2000), 1547–53; T. M. McKeever et al., "Early Exposure to Infections and Antibiotics and the Incidence of Allergic Disease," *Journal of Allergy and Clinical Immunology* 109 (2002), 43–50; Anne-Louise Ponsonby et al., "Relationship Between Early Life Respiratory Illness, Family Size over Time, and the Development of Asthma and Hay Fever: A Seven-Year Follow-up Study," *Thorax* 54 (1999), 664–69; Christine Stabell Ben et al., "Cohort Study of Sibling Effect, Infectious Diseases, and Risk of Atopic Dermatitis During the First 18 Months of Life," *British Medical Journal* 328 (2004), 1223–26.
23. David Poskanzer et al., "Polio and Multiple Sclerosis," *Lancet* (1963), 917–21; David Poskanzer et al., "Polio and Multiple Sclerosis," *Acta Neurological Scandinavica* 42-S19 (1966), 85–90.
24. C. C. Patterson et al., "Is Childhood-Onset Type 1 Diabetes a Wealth-Related Disease?" *Diabetologia* 44 (2001), supp 3:B9–16.
25. Patricia McKinney et al., "Early Social Mixing and Childhood Type 1 Diabetes Mellitus," *Diabetic Medicine* 17 (2000), 236–42.
26. A. L. Marshall et al., "Type 1 Diabetes Mellitus in Childhood: A Matched Case Control Study in Lancashire and Cumbria, U.K.," *Diabetic Medicine* 21 (2004), 1035–40.
27. Jean-Francois Bach, "The Effect of Infections on Susceptibility to Autoimmune and Allergic Diseases," *New England Journal of Medicine* 347 (2002), 911–20; EURODIAB ACE Study Group, "Variation and Trends in Incidence of Childhood Diabetes in Europe," *Lancet* 355 (2000), 873–76; F. Farrokhyar et al., "A Critical Review of Epidemiological Studies in Inflammatory Bowel Disease," *Scandinavian Journal of Gastroenterology* 36 (2001), 2–15.
28. American Academy of Allergy, Asthma and Immunology, *The Allergy Report: Science Based Findings on the Diagnosis and Treatment of Allergic Disorders, 1996–2001*; D. L. Jacobson et al., "Epidemiology and Estimated Population Bur-

den of Selected Autoimmune Diseases in the United States," *Clinical Immunology and Immunopathology* 84 (1997), 223–43.

29. Results of a membership survey of the Lupus Foundation of America, www.lupus .org.

30. Erika von Mutius et al. "Increasing Prevalence of Hay Fever and Atopy Among Children in Leipzig, East Germany," *Lancet* 351 (1998), 862–66.

31. Charlotte Braun-Fahrlander et al., "Prevalence of Hay Fever and Allergic Sensitization in Farmers' Children and Their Peers Living in the Same Rural Community," *Clinical and Experimental Allergy* 29 (1999), 28–34.

32. Josef Ridler et al., "Exposure to Farming in Early Life and Development of Asthma and Allergy: A Cross-Sectional Survey," *Lancet* 358 (2001), 1129–33; S. T. Remes et al., "Livestock over Field Crops: Which Factors Explain the Lower Prevalence of Atopy Amongst Farmers' Children," *Clinical and Experimental Allergy* 33 (2003), 427–34.

33. Charlotte Braun-Fahrlander, Erika von Mutius, and the Allergy and Endotoxin Study Team, "Environmental Exposure to Endotoxin and Its Relation to Asthma in School-Age Children," *New England Journal of Medicine* 347 (2002), 869–77.

34. "Dogs over Cats: Childhood Environment and Adult Atopy: Results from the European Community Respiratory Health Survey," *Journal of Allergy and Clinical Immunology* 103 (1999), 415–20; Marco Waser et al., "Exposure to Pets, and the Association with Hay Fever, Asthma, and Atopic Sensitization in Rural Children," *Allergy* 60 (2005), 177–84; Erika von Mutius et al., "Exposure to Endotoxin or Other Bacterial Components Might Protect Against the Development of Atopy," *Clinical and Experimental Allergy* 30 (2000), 1230–34.

35. Paolo Matricardi et al., "Exposure to Foodborne and Orofecal Microbes Versus Airborne Viruses in Relation to Atopy and Allergic Asthma: Epidemiological Study," *British Medical Journal* 320 (2000), 412–17. In an intriguing aside, in 2003 the American immunologist Dale Umetsu showed that hepatitis A, the only virus that has ever been associated with reduced allergy and asthma, may literally kill off the type-2 T cells that tend to drive these disorders. Dale Umetsu et al., "Hepatitis A Virus Link to Atopic Disease," *Nature* 425 (2003), 576.

36. Dale Umetsu et al., "Functional Heterogeneity Among Human Inducer T Cell Clones," *Journal of Immunology* 140 (1988), 4211–16.

37. Peter Yeung et al., "Heat-Killed *Listeria monocytogenes* as an Adjuvant Converts Established Murine Th2-Dominated Immune Responses into Th1-Dominated Responses," *Journal of Immunology* 151 (1998), 4146–52.

38. Gesine Hansen et al., "Allergen-Specific Th1 Cells Fail to Counterbalance Th2 Cell-Induced Airway Hyperreactivity but Cause Severe Airway Inflammation," *Journal of Clinical Investigation* 103 (1999), 175–83.

39. Philippe Stock et al., "Induction of T Helper Type 1-like Regulatory Cells That Express Foxp2 and Protect Against Airway Hyper-reactivity," *Nature Immunology* 5 (2004), 1149–56.

40. Shohei Hori et al., "Control of Regulatory T Cell Development by the Transcription Factor Foxp3," *Science* 299 (2003), 1057–61.

41. R. S. Wildin et al., "Clinical and Molecular Features of the Immunodysregulation, Polyendocrinopathy, Enteropathy, X Linked (IPEX) Syndrome," *Journal of Medical Genetics* 39 (2002), 537–45.

42. Stock et al., "Induction of T Helper Type 1-like Regulatory Cells."

43. Toll-like receptors get their name, not from any reference to toll booths, but from their similarity to a protein in fruit flies that, when defective, makes the flies look weird. The German word for "weird" is *toll*, and that was what Christiane Nusslein-Volhard, the German scientist who discovered this protein, called it.

44. This vital game of show-and-tell takes place primarily in the clumps of specialized immune tissue known as lymph nodes, which literally swell with the frenzied activity during an infection.

45. R. Maldonado-Lopez et al., "CD8 Subclasses of Dendritic Cells Direct the Development of Distinct T Helper Cells In Vivo," *Journal of Experimental Medicine* 189 (1999), 587–92.

46. Akiko Iwasaki and Ruslan Medzhitov, "Toll-like Receptor Control of the Adaptive Immune Responses," *Nature Immunology* 5 (2004), 987–95.

47. Daniel Hawiger et al., "Dendritic Cells Induce Peripheral T-cell Unresponsiveness Under Steady State Conditions In Vivo," *Journal of Experimental Medicine* 194 (2001), 769–79.

48. Ralph Steinman et al., "Tolerogenic Dendritic Cells," *Annual Review of Immunology* 21 (2003), 685–711.

49. Ralph Steinman et al., "Dendritic Cells Function In Vivo During the Steady State: A Role in Peripheral Tolerance," *Annals of the New York Academy of Sciences* 987 (2003), 15–27.

50. Stock et al., "Induction of T Helper Type 1-like Regulatory Cells."

51. Holden Maecker et al., "Vaccination with Allergin-IL-18 Fusion DNA Protects Against, and Reverses Established, Airway Hyperreactivity in a Murine Asthma Model," *Journal of Immunology* 166 (2001), 959–65.

52. Braun-Fahrlander et al., "Environmental Exposure to Endotoxin."

53. Waltraud Eder et al., "Toll-like Receptor 2 as a Major Gene for Asthma in Children of European Farmers," *Journal of Allergy and Clinical Immunology* 113 (2004), 482–88.

54. John M. Grange, "Effective Vaccination Against Tuberculosis—A New Ray of Hope," *Clinical and Experimental Immunology* 120 (2000), 232–34.

55. Christian Lienhardt and Alimuddin Zumla, "BCG: The Story Continues," *Lancet* 366 (2005), 1414–16.

56. G. M. Bahr et al., "Two Potential Improvements to BCG and Their Effect on Skin Test Reactivity in the Lebanon," *Tubercle* 67 (1986), 205–16; John Stanford, "Improving on BCG," *Acta Pathologica, Microbiologica et Immunologica* 99 (1991), 103–13.

57. G. D. Prema et al., "A Preliminary Report on the Immunotherapy of Psoriasis," *Indian Medical Gazette* 124 (1990), 381–82.

58. G. B. Marks et al., "The Effect of Neonatal BCG Vaccination on Atopy and Asthma at Age 7 to 14 Years: An Historical Cohort Study in a Community with a Very Low Prevalence of Tuberculosis Infection and a High Prevalence of Atopic Disease," *Journal of Allergy and Clinical Immunology* 111 (2003), 541–49; C. B. Sanjeevi et al., "BCG Vaccination and GAD65 and IA-2 Autoantibodies in Autoimmune Diabetes in Southern India," *Annals of the New York Academy of Sciences* 958 (2002), 293–96; Ramesh Bhonde and Pradeep Prab, "Can We Say Bye to BCG?" *Current Science* 77 (1999), 1283.

59. A. B. Alexandroff et al., "BCG Immunotherapy of Bladder Cancer: 20 Years On," *Lancet* 9165 (1999), 1689–94.

60. A 1999 review of the medical literature suggests that the Coley vaccine achieved survival rates comparable to that of oncologists treating the same types of tumors with modern methods. Stephen Hoption Cann et al., "Dr. William Coley and Tumour Regression, a Place in History or in the Future," *Postgraduate Medicine* 79 (2003), 672–80.

61. Mary Ann Richardson et al., "Coley Toxins Immunotherapy, a Retrospective Review," *Alternative Therapies in Health and Medicine* 5 (1999), 42–47.

62. *Allergy and asthma*: L. Camporota et al., "Effects of Intradermal Injection of SRL 172 on Allergen-Induced Airway Responses and IL-5 Generation by PBMC in Asthma," *Respiratory and Critical Care Medicine* 161 (2000); Peter Arkwright et al., "Intradermal Administration of a Killed *Mycobacterium vaccae* Suspension (SRL 172) Is Associated with Improvement in Atopic Dermatitis in Children with Moderate-to-Severe Disease," *Journal of Allergy and Clinical Immunology* 107 (2001), 531–34; L. Camporota et al., "The Effects of *Mycobacterium vaccae* on Allergen-Induced Airway Responses in Atopic Asthma," *European Respiratory Journal* 21 (2003), 287–93. *Cancer*: L. Assersohn et al., "A Randomised Pilot Study of SRL 172 (*Mycobacterium vaccae*) in Patients with Small Cell Lung Cancer Treated with Chemotherapy," *Clinical Oncology* 14 (2002), 23–27; R. Mendes et al., "Clinical and Immunological Assessment of *Mycobacterium vaccae* (SRL 172) with Chemotherapy in Patients with Malignant Mesethelioma," *British Journal of Cancer* 86 (2002), 336–41; S. Nicholson et al., "A Randomized Phase II Trials of SRL 172 (*Mycobacterium vaccae*) +/– Low-Dose Interleukin-2 in the Treatment of Metastatic Malignant Melanoma," *Melanoma Research* 13 (2003), 389–93; A. Maraveyas et al., "Possible Improved Survival of Patients with Stage IV AJCC Melanoma Receiving SRL 172 Immunotherapy: Correlation with Induction of Increased Levels of Intracellular Interleukin-2 in Peripheral Blood Lymphocytes," *Annals of Oncology* 10 (1999), 817–24; John Stanford et al., "A Phase II Single Arm Trials of *Mycobacterium vaccae* (SRL172) Monotherapy in Renal Cell Carcinoma Compared with Historical Controls," *European Cancer Journal* 2006 (prepublication manuscript).

63. *The Dirt Vaccine* (airing BBC 2002), program in *The Edge Series: Health and Medicine*, Eco Services Film; Garry Hamilton, "Let Them Eat Dirt," *New Scientist*, July 18, 1998; Susan McCarthy, "Talking Dirty and Bring on the Germs," salon.com, April 2000.

64. Mary O'Brien et al., "SRL172 (Killed *Mycobacterium vaccae*) in Addition to Standard Chemotherapy Improves Quality of Life Without Affecting Survival, in Patients with Advanced Non-small-cell Lung Cancer: Phase III Results," *Annals of Oncology* 15 (2004), 906–14.

65. John Stanford et al., "Successful Immunotherapy with *Mycobacterium vaccae* in the Treatment of Adenocarcinoma of the Lung," unpublished.

66. On April 11, 2001, SR Pharma shares dropped 76.9 percent, the biggest drop on the FTSE 100 Index that day.

67. National Sweet Itch Centre, "Studies on the Prevention and Treatment of Sweet-Itch," www.sweet-itch.co.uk/trials.html.

68. John Stanford and Graham McIntyre, unpublished results, Centre for Infectious Diseases and International Health, University College London, 2006.

69. Claudia Zuany-Amorin et al., "Suppression of Airway Eosinophilia by Killed *My-

cobacterium vaccae–Induced Allergen-Specific Regulatory T-cells," *Nature Medicine* 8 (2002), 625–29; Victoria Adams et al., "*Mycobacterium vaccae* Induces a Population of Pulmonary Cd11c Antigen Presenting Cells That Secrete Regulatory Cytokines," *European Journal of Immunology* 34 (2004), 631–38.
70. Graham Rook, unpublished results, February 2006.
71. Fawziah Marra et al., "Does Antibiotic Exposure During Infancy Lead to Development of Asthma? A Systematic Review and Meta-analysis (of Seven Studies)," *Chest* 129 (2006), 610–18.
72. Mairi Noverr et al., "Development of Allergic Airway Disease in Mice Following Antibiotic Therapy and Fungal Microbiota Increase: Role of Host Genetics, Antigen, and Interleukin-13," *Infection and Immunity* 73 (2005), 30–38; Mairi Noverr and Gary Huffnagle, "Does the Microbiota Regulate Immune Responses Outside the Gut?" *Trends in Microbiology* 12 (2004), 562–68; Mairi Noverr and Gary Huffnagle, "Role of Antibiotics and Fungal Microbiota in Driving Pulmonary Allergic Responses," *Infection and Immunity* 72 (2004), 4996–5003.
73. B. Laubereau et al., "Caesarean Section and Gastrointestinal Symptoms, Atopic Dermatitis, and Sensitisation During the First Year of Life," *Archives of Diseases of Children* 89 (2004), 993–97.
74. A. C. Ouwehand et al., "Differences in *Bifidobacterium* Flora Composition in Allergic and Healthy Infants," *Journal of Allergy and Clinical Immunology* 108 (2001), 144–45.
75. B. Björksten et al., "The Intestinal Microflora in Allergic Estonian and Swedish Infants," *Clinical and Experimental Allergy* 29 (1999), 342–46; E. Sepp et al., "Intestinal Microflora of Estonian and Swedish Infants," *Acta Paediatrica* 86 (1997), 956–61.
76. Anita van den Biggelaar et al., "Long-Term Treatment of Intestinal Helminths Increases Mite Skin-Test Reactivity in Gabonese Schoolchildren," *Journal of Infectious Diseases* 189 (2004), 892–900.
77. R. W. Summers et al., "*Trichuris suis* Therapy in Crohn's Disease," *Gut* 54 (2005), 87–90.
78. Charlotte Schubert, "The Worm Has Turned," *Nature Medicine* 10 (2004), 1204–71.
79. Christopher Lowry, "Functional Subsets of Serotonergic Neurones," *Journal of Neuroendocrinology* 14 (2002), 911–23; Christopher Lowry et al., "Modulation of Anxiety Circuits by Serotonergic Systems," *Stress* 8 (2005), 233–46.
80. Rose-Marie Bluthe et al., "Central Injection of IL-10 Antagonizes the Behavioral Effects of Lipopolysaccharide in Rats," *Psychoneuroendocrinology* 24 (1999), 301–11.
81. Marianne Wamboldt et al., "Familial Association Between Allergic Disorders and Depression in Adult Finnish Twins," *American Journal of Medical Genetics* 96 (2000), 146–53.
82. M. Maes et al., "In Vitro Immunoregulatory Effects of Lithium in Healthy Volunteers," *Psychopharmacology* 143 (1999), 401; Marta Kubera et al., "Anti-inflammatory Effects of Antidepressants Through Suppression of the Interferon-[gamma] Interleukin-10 Production Ratio," *Journal of Clinical Psychopharmacology* 21 (2001), 199–206; Brian Leonard, "The Immune System, Depression and the Action of Antidepressants," *Progress in Neuro-Psychopharmacology and Biological Psychiatry* 25 (2001), 767–80; Sinead O'Brien et al., "Cytokines: Abnormalities in Major Depression and Implications for Pharmacological Treatment," *Human*

Psychopharmacology: Clinical and Experimental 19 (2004), 397–403; Nathalie Castanon et al., "Chronic Administration of Tianeptine Balances Lipopolysaccharide-Induced Expression of Cytokines in the Spleen and Hypothalamus of Rats," *Psychoneuroendocrinology* 29 (2004), 778–90.

PART IV: BUGS ON DRUGS

Part opener: Tom Nugent, "Resistance Fighter," *Pennsylvania Gazette*, September–October 2000.

1. "New Germ Strain Takes Heavy Toll," *New York Times*, March 22, 1958, 19.
2. Studies show 20 to 30 percent of us to be "persistent carriers," with one particular strain of *Staph. aureus.* showing up inside our nostrils on a consistent basis. Another 50 to 60 percent of us periodically pick up and lose a succession of different strains over time. And a lucky 10 to 20 percent remain staph-free. Research further suggests that the variation stems from inherent differences in the immune chemicals in our nasal secretions. See J. Kluytmans et al., "Nasal Carriage of *Staphylococcus aureus*: Epidemiology, Underlying Mechanisms, and Associated Risks," *Clinical Microbiology Reviews* 10 (1997), 505–20.
3. In 1957 and 1958 U.S. infant mortality increased for the first time since 1936, from 108,000 deaths (26.0 per 1,000 live births) in 1956, to 112,000 (26.3 per 1,000 births) in 1957 and 114,000 deaths (27.1 per 1,000 births) in 1958. They inched back down to below 1956 levels in 1961 (25.3 per 1,000 live births). U.S. Census Bureau, *Live Births, Deaths, Infant Deaths, and Maternal Deaths: 1900 to 2001*, Statistical Abstract No. HS-13. (No statistics on specific cause by organism.)
4. M. Barber, "Methicillin-Resistant Staphylococci," *Journal of Clinical Pathology* 1 (1963), 308–11.
5. Heinz Eichenwald, interviews by author, January 2005 and August 2006.
6. With a few strange exceptions, all bacteria have one of two main types of cell wall, identified with a microscopic staining technique developed by the nineteenth-century Danish microbiologist Hans Christian Gram. In essence, gram-positive bacteria are encased by a thick layer of the meshlike molecule peptidoglycan. Gram-negative bacteria have a much thinner peptidoglycan cell wall, but it is encapsulated by an additional outer membrane made of lipopolysaccharide and phospholipid.
7. Joshua Lederberg and Esther Lederberg, "Replica Plating and Indirect Selection of Bacterial Mutants," *Journal of Bacteriology* 83 (1952), 399–406.
8. Edward Tatum, "Gene Recombination in *Escherichia coli*," Cold Spring Harbor Symposia, June 1946; Edward Tatum and Joshua Lederberg, "Gene Recombination in the Bacterium *Escherichia coli*," *Journal of Bacteriology* 53 (1947), 673–84.
9. Joshua Lederberg, Luigi Cavalli, and Esther Lederberg, "Sexual Compatibility in *Escherichia coli*," *Genetics* 37 (1952), 720–30.
10. Lederberg's term *plasmid* for a set of nonchromosomal bacterial genes paralleled the already coined term *plasmagene*, which refers to the nonchromosomal genes found outside the nucleus, or in the cyto*plasm*, of the cells of all higher animals and plants. See Joshua Lederberg, "Cell Genetics and Hereditary Symbiosis," *Physiological Review* 32 (1952), 403–26.

11. Esther Lederberg and Joshua Lederberg, "Genetic Studies of Lysongenicity in *Escherichia coli*," *Genetics* 38 (1953), 51–64.

12. *Lambda* would go on to become the workhorse of genetic engineering, used to create transgenic life by picking up the genes from one organism and splicing them into another.

13. Norton Zinder and Joshua Lederberg, "Infective Heredity in Bacteria," *Journal of Bacteriology* 64 (1952), 679–99.

14. F Griffith, "The Significance of Pneumococcal Types," *Journal of Hygiene.* 27 (1928), 113–59.

15. Oswald Avery et al., "Studies on the Chemical Nature of the Substance Inducing Transformation of Pneumococcal Types," *Journal of Experimental Medicine* 83 (1944), 89–96.

16. In 1958 Joshua Lederberg shared a Nobel Prize with George Beadle and Edward Tatum for his discoveries concerning gene recombination, or "sex," in bacteria. The Lederbergs divorced in 1968. Joshua went on to become the president of Rockefeller University in New York City in 1978; Esther remained a professor at Stanford University, where she maintained the world's most extensive collection of bacterial plasmids until she suffered a debilitating heart attack in 1999. Unfortunately, historic accounts have largely glossed over Esther's vital role in the couple's historic discoveries. Both Lederbergs granted me personal interviews for this book.

17. T. Watanabe and T. Fukasawa, "Episome-Mediated Transfer of Drug Resistance in Enterobacteriacea," *Journal of Bacteriology* 81 (1961), 669–78.

18. K. Ochiai et al., "Studies on Inheritance of Drug Resistance Between *Shigella* Strains and *Escherichia coli* Strains," *Nihon Iji Shimpo* 1861 (1959), 34–46 (in Japanese), and K. Ochiai et al., "Studies on the Mechanism of Development of Multiple Drug Resistant *Shigella* Strains," *Nihon Iji Shimpo* 1866 (1960), 45–50 (in Japanese), as reported in Watanabe and Fukasawa, "Episome-Mediated Transfer."

19. In 2001 researchers discovered a cluster of drug-resistance genes in the midst of a so-called "pathogenicity island," a large transposon conveying a variety of virulence traits such as dramatically increased toxin production. See S. N. Luck et al., "Ferric Dicitrate Transport System of *Shigella Flexneri 2a YSH6000* Is Encoded in a Novel Pathogenicity Island Carrying Multiple Antibiotic Resistance Genes," *Infection and Immunology* 69 (2001), 6012–21.

20. L. M. Mundy et al., "Relationship Between Enterococcal Virulence and Antimicrobial Resistance," *Clinical Microbiology Reviews* (October 2000), 513–22.

21. Robert Weinstein, "Nosocomial Infection Update," *Emerging Infectious Diseases* 4 (1998), 416–20.

22. *ASM News*, June 2004 (American Society for Microbiology).

23. R. Leclercq et al., "Plasmid-Mediated Resistance to Vancomycin and Teicoplanin in *Enterococcus faecium*," *New England Journal of Medicine* 319 (1988), 157–61.

24. "Nosocomial Enterococci Resistant to Vancomycin—United States, 1989–1993," *Morbidity and Mortality Weekly Report* 42 (1993), 597–99.

25. "*Staphylococcus aureus* Resistant to Vancomycin—United States 2002," *Morbidity and Mortality Weekly Report*, July 5, 2002.

26. "Brief Report: Vancomycin-Resistant *Staphylococcus aureus*—New York 2004," *Morbidity and Mortality Weekly Report*, April 23, 2004.

27. Oxazolidinones prevent the bacterial ribosome from initiating protein synthesis by blocking the attachment of the messenger RNA molecules that spell out the instructions for assembling proteins one amino acid at a time.

28. Neil Woodford et al., "Detection of Oxazolidinone-Resistant *Enterococcus faecalis* and *Enterococcus faecium* Strains by Real-Time PCR and PCR-Restriction Fragment Length Polymorphism Analysis," *Journal of Clinical Microbiology* 40 (2002), 4298–300.

29. R. Devasia et al., "The First Reported Hospital Outbreak of Linezolid-Resistant Enterococcus: An Infection Control Problem Has Emerged," Infectious Disease Society of America Meeting 2005, abstract 1079.

30. Curtis Donskey et al., "Effect of Parenteral Antibiotic Administration on Persistence of Vancomycin-Resistant *Enterococcus faecium* in the Mouse Gastrointestinal Tract," *Clinical Infectious Disease* 180 (1999), 384–90.

31. Curtis Donskey et al., "Effect of Antibiotic Therapy on the Density of Vancomycin-Resistant Enterococci in the Stool of Colonized Patients," *New England Journal of Medicine* 343 (2000), 1925–32.

32. Michelle Hecker et al., "Unnecessary Use of Antimicrobials in Hospitalized Patients: Current Patterns of Misuse with an Emphasis on the Antianaerobic Spectrum of Activity," *Archives of Internal Medicine* 163 (2003), 972–78.

33. "Severe *Clostridium difficile*–Associated Disease in Populations Previously at Low Risk—Four States, 2005," *Morbidity and Mortality Weekly*, December 2, 2005.

34. R. Viscidi et al., "Isolation Rates and Toxigenic Potential of *Clostridium difficile* Isolates from Various Patient Populations," *Gastroenterology* 81 (1981), 5–9; L. V. McFarland et al., "Nosocomial Acquisition of *Clostridium difficile* Infection," *New England Journal of Medicine* 320 (1989), 204–10.

35. Paul Byrne, "Toenail Surgery Nearly Killed Me, Jamie-Lee, 15, One of Youngest Victims," *Mirror* [U.K.], August 30, 2005; "Hospital Blamed for Mum's Horrible Death," *Windsor* [U.K.] *Express*, March 31, 2006.

36. Vivian Loo et al., "A Predominantly Clonal Multi-Institutional Outbreak of *Clostridium difficile*–Associated Diarrhea with High Morbidity and Mortality," *New England Journal of Medicine* 353 (2005), 2442–49.

37. L. Clifford McDonald et al., "An Epidemic, Toxin Gene-Variant Strain of *Clostridium difficile*," *New England Journal of Medicine* 353 (2005), 2433–41.

38. Carlene Muto et al., "A Large Outbreak of *Clostridium difficile*–Associated Disease with an Unexpected Proportion of Deaths and Colectomies at a Teaching Hospital Following Increased Fluoroquinolone Use," *Infection Control Hospital Epidemiology* 3 (2005), 273–80.

39. Luis Fabregas, "Superbug Infecting Area Patients," *Pittsburgh Tribune-Review*, October 29, 2005.

40. "Severe *Clostridium difficile* Associated Disease in Populations Previously at Low Risk—Four States, 2005," *Morbidity and Mortality Weekly Report*, December 2, 2005.

41. Yves Gillet et al., "Association Between *Staphylococcus aureus* Strains Carrying Gene for Panton-Valentine Leukocidin and Highly Lethal Necrotising Pneumonia in Young Immunocompetent Patients," *Lancet* 359 (2002), 753–59.

42. Isaac Starr, third-year medical student, University of Pennsylvania, 1918, as quoted

in "Influenza in 1918: Recollections of the Epidemic in Philadelphia," *Annals of Internal Medicine* 145 (2006), 138–40.

43. Betsy Herold et al., "Community-Acquired Methicillin-Resistant *Staphylococcus aureus* in Children with No Identified Predisposing Risk," *Journal of the American Medical Association* 279 (1998), 593–98.

44. "Four Pediatric Deaths from Community-Acquired Methicillin-Resistant *Staphylococcus aureus*—Minnesota and North Dakota, 1997–1999," *Morbidity and Mortality Weekly Report*, August 20, 1999, 707–10.

45. Carlos Sattler et al., "Prospective Comparison of Risk Factors and Demographic and Clinical Characteristics of Community-Acquired, Methicillin-Resistant versus Methicillin-Susceptible *Staphylococcus aureus* Infection in Children," *Pediatric Infectious Disease Journal* 21 (2002), 910–16.

46. Sophia Kazakova et al., "A Clone of Methicillin-Resistant *Staphylococcus aureus* Among Professional Football Players," *New England Journal of Medicine* 352 (2005), 468–75.

47. D. A. Robinson et al., "Re-emergence of Early Pandemic *Staphylococcus aureus* as a Community-Acquired Methicillin-Resistant Clone," *Lancet* 365 (2005), 1256–58.

48. Sheldon Kaplan et al., "Three-year Surveillance of Community-Acquired *Staphylococcus aureus* Infections in Children," *Clinical Infectious Diseases* 40 (2005), 1785–91; M. D. King et al., "Emergence of Community-Acquired Methicillin-Resistant *Staphylococcus aureus* USA 300 Clone as the Predominant Cause of Skin and Soft-Tissue Infections," *Annals of Internal Medicine* 144 (2006), 309–17; Gregory Moran et al., "Methicillin-Resistant *S. aureus* Infections Among Patients in the Emergency Department," *New England Journal of Medicine* 355 (2006), 666–74.

49. Tsutomu Watanabe et al., "Episome-Mediated Transfer of Drug Resistance in Enterobacteriaceae X," *Journal of Bacteriology* 92 (1966), 477–486.

50. Ellen C. Moorhouse, "Transferable Drug Resistance in Enterobacteria Isolated from Urban Infants," *British Medical Journal* 2 (1969), 405.

51. K. B. Linton et al., "Antibiotic Resistance and Transmissible R-factors in the Intestinal Coliform Flora of Healthy Adults and Children in an Urban and Rural Community," *Journal of Hygiene* 70 (1972), 99–104.

52. D. V. Sompolinsky et al., "Microbiological Changes in the Human Fecal Flora Following the Administration of Tetracyclines and Chloramphenicol," *American Journal of Proctology* 18 (1967), 471–78.

53. Stuart Levy et al., "High Frequency of Antimicrobial Resistance in Human Fecal Flora," *Antimicrobial Agents and Chemotherapy* 32 (1988), 1801–6.

54. Stuart Levy, interview by author, June 2006.

55. N. B. Shoemaker, H. Hayes, and A. A. Salyers, "Evidence for Extensive Resistance Gene Transfer Among Bacteroides spp. and Among Bacteroides and Other Genera in the Human Colon," *Applied and Environmental Microbiology* 67 (February 2001), 561–68.

56. A species name that many bacteriologists rank among the worst of all time, *thetaiotaomicron* is a combination of the three Greek letters (theta, iota, and omicron) that the Romanian bacteriologist A. Distaso thought he saw when looking down the barrel of his microscope in 1912 to discover the jumble of short and long

rods that grew out from this, his newly discovered bacterial species. A. Distaso, *Zentralblatt fur Bakteriologie, Parasitenkunde, Infektionskrankheiten, und Hygiene*. Abteilung I. 62 (1912) 433–68. "What is lost to history," Salyers adds, "is whether this naming inspiration came after a visit to the nearby pub." For good reason, most microbiologists who work with this species refer to it as *B. theta* outside their formal writings.

57. Laura McMurry et al., "Triclosan Targets Lipid Synthesis," *Nature* 394 (1998), 531–32; S. P. Cohen, H. Hachler, and S. B. Levy, "Genetic and Functional Analysis of the Multiple Antibiotic Resistant (*mar*) Locus in *Escherichia coli*," *Journal of Bacteriology* 175 (1993), 1484–92; M. C. Moken, L. M. McMurry, and S. B. Levy, "Selection of Multiple Antibiotic Resistant (*mar*) Mutants of *Escherichia coli* by Using the Disinfectant Pine Oil," *Antimicrobial Agents and Chemotherapy* 41 (1997), 2770–72; L. M. McMurry, M. Oethinger, and S. B. Levy, "Overexpression of marA, soxS, or acrAB Produces Resistance to Triclosan in *Esherichia coli*," *FEMS Microbiology Letters* 166 (1998), 305–9.

58. Rolf Halden et al., "Co-occurrence of Triclocarban and Triclosan in U.S. Water Resources," *Environmental Science and Technology* 39 (2005), 1420–26; Jochen Heidler et al., "Partitioning, Persistence and Accumulation in Digested Sludge of the Topical Antiseptic Triclocarban During Wastewater Treatment," *Environmental Science and Technology* 40 (2006), 3634–39.

59. Holli Lancaster et al., "Prevalence and Identification of Tetracycline-Resistant Oral Bacteria in Children Not Receiving Antibiotic Therapy," *FEMS Microbiology Letters* 228 (2003), 99–104.

60. The idea that antibiotics evolved at least in part as signaling is that of microbiologist Julian Davies, at the University of British Columbia, in Vancouver. His lab's seminal paper on the subject is Ee-Been Goh et al., "Transcriptional Modulation of Bacterial Gene Expression by Subinhibitory Concentrations of Antibiotics," *Proceedings of the National Academy of Sciences* 99 (2002), 17025–30.

61. C. G. Marshall et al., "Glycopeptide Antibiotic Resistance Genes in Glycopeptide-Producing Organisms," *Antimicrobial Agents and Chemotherapy* 42 (1998), 2215–20.

62. The chemical name for Synercid is quinupristin-dalfopristin; that for Tygacil is tigecycline; and that for Cubicin is daptomycin.

63. The chemical name for Ketek is telithromycin, and that for Zyvox is linezolid.

64. Vanessa D'Costa et al., "Sampling the Antibiotic Resistome," *Science* 311 (2006), 374–77.

65. "Superbugs Abound in Soil," www.nature.com/news/2006/060116/full/060116-10 .html.

66. William Laurence, "Wonder Drug Aureomycin Found to Spur Growth 50 Percent," *New York Times*, April 10, 1950, 1.

67. Notes on Science, *New York Times*, May 20, 1950, E9.

68. Animal Health Institute, "Antibiotic Use in Animals Rises in 2004," press release, June 27, 2005.

69. "Hogging It! Estimates of Antimicrobial Abuse in Livestock," Union of Concerned Scientists, January 2001, executive summary, xiii, www.ucsusa.org/food_and_ environment/antibiotics_and_food/hogging-it-estimates-of-antimicrobial-abuse-in-livestock.html.

70. Animal Health Institute, "Antibiotic Use in Animals Rises in 2004," press release, June 27, 2005.

71. Animal Health Institute, *Active Antibacterial Ingredients Sold by AHI Members, 2002–2004 AHI Survey*, June 27, 2005.

72. Amy Chapin et al., "Airborne Multidrug-Resistant Bacteria Isolated from a Concentrated Swine Feeding Operation," *Environmental Health Perspectives* 113 (2005), 137; M. P. Schlusener and K. Bester, "Persistence of Antibiotics Such As Macrolides, Tiamulin and Salinomycin in Soil," *Environmental Pollution* February 2, 2006 [Epub ahead of print]; J. M. Cha et al., "Rapid Analysis of Trace Levels of Antibiotic Polyether Ionophores in Surface Water by Solid-Phase Extraction," *Journal of Chromatography* 1065 (2005), 187–98.

73. Morten Helms et al., "Excess Mortality Associated with Antimicrobial Drug-Resistant *Salmonella typhimurium*," *Emerging Infectious Diseases* 8 (2002), 490–95; J. K. Varma et al., "Antimicrobial Resistance in Salmonella Is Associated with Increased Hospitalizations," presented at National Antibiotic Resistance Monitoring System 1996–2000, International Conference on Emerging Infectious Diseases, 2002.

74. H. A. Elder et al., "Human Studies to Measure the Effect of Antibiotic Residues," *Veterinary and Human Toxicology* 35 (1996), suppl. 1, 31–36.

75. Jeffrey LeJeune and Nicholas Christie, "Microbiological Quality of Ground Beef from Conventionally Reared Cattle and 'Raised Without Antibiotics' Label Claims," *Journal of Food Protection*, 67 (2004), 1433–37.

76. M. Meyer et al., "Occurrence of Antibiotics in Surface and Ground Water near Confined Animal Feeding Operations and Waste Water Treatment Plants Using Radioimmunoassay and Liquid Chromatography/Electrospray Mass Spectrometry," U.S. Geological Survey, Raleigh, N.C.

77. Amee Manges et al., "Widespread Distribution of Urinary Tract Infections Caused by a Multidrug-Resistant *Escerichia coli* Clonal Group," *New England Journal of Medicine* 345 (2001), 1007–13; "An Epidemic of Urinary Tract Infections?" editorial, *New England Journal of Medicine* 345 (2001), 1055–57.

78. M. M. Swann, *Report of the Joint Committee on the Use of Antibiotics in Animal Husbandry and Veterinary Medicine* (London: Her Majesty's Stationery Office, 1969).

79. Henrick Wegener, "Ending the Use of Antimicrobial Growth Promoters Is Making a Difference: In Denmark Antibiotic Resistance Levels Fell While Food Productivity Remains Strong," *American Society of Microbiology News* 69 (2003), 443–48.

80. Ingo Klare et al., "Occurrence and Spread of Antibiotic Resistances in *Enterococcus faecium*," *International Journal of Food Microbiology* 88 (2003), 269–90.

81. F. Angulo et al., "Isolation of Quinupristin/Dalfopristin-Resistant *Enterococcus faecium* from Human Stool Specimens and Retail Chicken Products in the United States," presentation at the First International Conference on Enterococci, Banff, Canada, February 2000.

82. Amy Kieke et al., "Use of Streptogramin Growth Promoters in Poultry and Isolation of Streptogramin-Resistant *Enterococcus faecium* from Humans," *Journal of Infectious Diseases* 194 (2006), 1200–1208.

PART V: FIGHTING SMARTER, NOT HARDER

1. *Aging of Veterans of the Union Army: Military, Pension and Medical Records, 1820–1940* (ICPSR 6837); A. J. Bollet "Rheumatic Diseases Among Civil War Troops," *Arthritis and Rheumatism* 34 (1991), 1197–1203; Gina Kolata, "So Big and Healthy Nowadays, Grandpa Wouldn't Know You," *New York Times*, July 30, 2006, A1.

2. Robert Fogel and Dora Costa, "A Theory of Technophysio Evolution, with Some Implications for Forecasting Population, Health Care Costs, and Pension Costs," *Demography* 34 (1997), 49–66; Dora Costa, "Understanding the 20th-Century Decline in Chronic Conditions Among Older Men," *Demography* 37 (2000), 53–72; Dora Costa, "Why Were Older Men in the Past in Such Poor Health," available online at web.mit.edu/costa/www/papers.html.

3. Jianhui Zhu et al., "Prospective Study of Pathogen Burden and Risk of Myocardial Infarction or Death," *Circulation* 103 (2001), 45–51; Jean-Louis Georges et al., "Impact of Pathogen Burden in Patients with Coronary Artery Disease in Relation to Systemic Inflammation and Variation in Genes Encoding Cytokines," *American Journal of Cardiology* 92 (2003), 515–21.

4. Caleb Finch and Eileen Crimmins, "Inflammatory Exposure and Historical Changes in Human Life-Spans," *Science* 305 (2004), 1736–39; Eileen Crimmins and Caleb Finch, "Infection, Inflammation, Height and Longevity," *Proceedings of the National Academy of Sciences* 103 (2006), 498–503.

5. R. Montenegro and C. Stephens, "Indigenous Health in Latin America and the Caribbean," *Lancet* 367 (2006), 1859–69.

6. Rudi Westendorp et al., "Optimizing Human Fertility and Survival," *Nature Medicine* 7 (2001), 873; G. Doblhammer and J. Oeppen, "Reproduction and Longevity: The Effect of Frailty and Health Selection," *Proceedings of the Royal Society of London* 270 (2003), 1541–47; D. Lio et al., "Inflammation, Genetics, and Longevity," *Journal of Medical Genetics* 40 (2003), 296–99.

7. Rudi Westendorp, "Are We Becoming Less Disposable?" *EMBO Reports* 5 (2004), 2–6.

8. Y. W. Miller et al., "Sequential Antibiotic Therapy for Acne Promotes the Carriage of Resistant Staphylococci on the Skin of Contacts," *Journal of Antimicrobial Chemotherapy* 38 (1996), 829–37; David Margolis et al., "Antibiotic Treament of Acne May Be Associated with Upper Respiratory Tract Infections," *Archives of Dermatology* 141 (2005), 1132–36; Ross Levy et al., "Effect of Antibiotics on the Oropharyngeal Flora in Patients with Acne," *Archives of Dermatology* 139 (2003), 467–71.

9. K. G. Naber, "Treatment Options for Acute Uncomplicated Cystitis in Adults," *Journal of Antimicrobial Chemotherapy* 46 (2000), S23–S277.

10. Susan Swedo et al., "Pediatric Autoimmune Neuropsychiatric Disorders Associated with Streptococcal Infections: Clinical Description of the First 50 Cases," *American Journal of Psychiatry* 155 (1998), 264–71; F. Breedveld et al., "Minocycline Treatment for Rheumatoid Arthritis: An Open Dose Study," *Journal of Rheumatology* 17 (1990), 43–46.

11. Marie-Therese Labro, "Antibiotics as Anti-inflammatory Agents," *Current Opinions in Investigational Drugs* 3 (2002), 61–68; P. N. Black, "Anti-inflammatory Effects of Macrolide Antibiotics," *European Respiratory Journal* 10 (1997), 971–72.

12. J. Thomas Grayston, "Does *Chlamydia pneumoniae* Cause Atherosclerosis?" *Archives of Surgery* 134 (1999), 930–34; J. D. Beck et al., "Dental Infections and Atherosclerosis," *American Heart Journal* 138 (1999), S528–33.

13. Francisco Gimenez-Sanchez et al., "Treating Cardiovascular Disease with Antimicrobial Agents: A Survey of Knowledge, Attitudes, and Practices Among Physicians in the United States," *Clinical Infectious Diseases* 33 (2001), 171–76.

14. Christopher Cannon et al., "Antibiotic Treatment of *Chlamydia pneumoniae* After Acute Coronary Syndrome," *New England Journal of Medicine* 352 (2005), 1646–54; Thomas Grayston et al., "Azithromycin for the Secondary Prevention of Coronary Events," *New England Journal of Medicine* 352 (2005), 1637–45.

15. Jeffrey Anderson, "Infection, Antibiotics and Atherothrombosis—End of the Road or New Beginnings," *New England Journal of Medicine* 352 (April 21, 2005), 1706–9, available online at www.activbiotics.com.

16. *WHO Annual Report on Infectious Disease: Overcoming Antmicrobial Resistance* (Geneva: World Health Organization, 2000); available online at www.who.int/infectious-disease-report/2000/.

17. Amy Pruden et al., "Antibiotic Resistance Genes as Emerging Contaminants: Studies in Northern Colorado," *Environmental Science and Technology*, Web release August 15, 2006.

18. Elissa Ladd, "The Use of Antibiotics for Viral Upper Respiratory Tract Infections: An Analysis of Nurse Practitioner and Physician Prescribing Practices in Ambulatory Care, 1997–2001," *Journal of the American Academy of Nurse Practitioners* 17 (2005), 416–24; Arch Mainous III et al., "Trends in Antimicrobial Prescribing for Bronchitis and Upper Respiratory Infections Among Adults and Children," *American Journal of Public Health* 93 (2003), 1910–14.

19. H. Bauchner et al., "Parents, Physicians, and Antibiotic Use," *Pediatrics* 103 (1999), 395–98; R. L. Watson et al., "Antimicrobial Use for Pediatric Upper Respiratory Infections: Reported Practice, Actual Practice, and Parent Beliefs," *Pediatrics* 104 (1999), 1251–57; R. L. Watson et al., "Inappropriateness and Variability of Antibiotic Prescription Among French Office-Based Physicians," *Journal of Clinical Epidemiology* 51 (1998), 61–68; E.E.L. Wang et al., "Antibiotic Prescribing for Canadian Preschool Children: Evidence of Overprescribing for Viral Respiratory Infections," *Clinical Infectious Diseases* 29 (1999), 155–60; L. F. McCaig and J. M. Hughes, "Trends in Antimicrobial Drug Prescribing Among Office Based Physicians in the United States," *Journal of the American Medical Association* 273 (1995), 214–19.

20. Rachida el Moussaoui et al., "Effectiveness of Discontinuing Antibiotic Treatment After Three Days Versus Eight Days in Mild to Moderate-Severe Community Acquired Pneumonia: Randomised, Double-Blind Study," *British Medical Journal* 332 (2006), 1355–62.

21. Christopher Stille et al., "Increased Use of Second-Generation Macrolide Antibiotics for Children in Nine Health Plans in the United States," *Pediatrics* 114 (2004), 1206–11; Mainous et al., "Trends in Antimicrobial Prescribing."

22. William Check, "Real-Time PCR for the Rest of Us," College of American Pathologists/*CAP Today*, June 2006.

23. Didier Guillemot et al., "Reduction of Antibiotic Use in the Community Reduces the Rate of Colonization with Penicillin G-nonsusceptible *Streptococcus pneumoniae*," *Clinical Infectious Diseases* 41 (2005), 930–38.

24. W. Michael Dunne et al., "Clinical Microbiology in the Year 2025," *Journal of Clinical Microbiology* 40 (2002), 3889–93.
25. W. J. Wilson et al., "Sequence-Specific Identification of 18 Pathogenic Micro-organisms Using Microarray Technology," *Molecular and Cellular Probes* 16 (2002), 119–27.
26. Tom Slezak, correspondence to author, November 2006.
27. Daniel Sinsimer et al., "Use of Multiplex Molecular Beacon Platform for Rapid Detection of Methicillin and Vancomycin Resistance in *Staphylococcus aureus*," *Journal of Clinical Microbiology* 43 (2005), 4585–91.
28. J. T. McLure et al., "Performance of an Investigational Commercial Real-Time PCR Assay for Direct Detection of *Staphylococcus aureus* and MRSA on Clinical Samples," poster 357, 44th annual meeting of the Infectious Disease Society of America, Toronto, October 10, 2006.
29. Ibid.
30. Lori Henderson, GeneOhm associate director of marketing, correspondence to author, November 2006.
31. Jaana Harmoinen et al., "Orally Administered Targeted Recombinant Beta-lactamase Prevents Ampicillin-Induced Selective Pressure on the Gut Microbiota: A Novel Approach to Reducing Antimicrobial Resistance," *Antimicrobial Agents and Chemotherapy* 48 (2004), 75–79; "Ipsat Therapies Announces Positive Phase II Results for Lead Product Against Antibiotic Resistance and Hospital Acquired Infection," data presented at the 15th European Congress of Clinical Microbiology and Infectious Diseases, Copenhagen, April 5, 2005.
32. Klaus Stoeckel et al., "Stability of Cephalosporin Prodrug Esters in Human Intestinal Juice: Implications for Oral Bioavailability," *Antimicrobial Agents and Chemotherapy* 42 (1998), 2602–6.
33. W. Graninger, "Pivmecillinam—Therapy of Choice for Lower Urinary Tract Infection," *International Journal of Antimicrobial Agents* 22 (2003), suppl. 2:73–78; George Zhanel et al., "A Canadian National Surveillance Study of Urinary Tract Isolates from Outpatients: Comparison of the Activities of Trimethoprim-Sulfamethoxazole, Ampicillin, Mecillinam, Nitrofurantoin, and Ciprofloxacin," *Antimicrobial Agents and Chemotherapy* 44 (2000), 1089–92.
34. A. Sullivan et al., "Effect of Perorally Administered Pivmecillinam on the Normal Oropharyngeal, Intestinal and Skin Microflora," *Journal of Chemotherapy* 13 (2001), 299–308; A. Heimdahl et al., "Effect of Bacampicillin on Human Mouth, Throat and Colon Microflora," *Infection* 7 (1979), S446–51.
35. Carolos Amabile-Cuevas and Jack Heinemann, "Shooting the Messenger of Antibiotic Resistance: Plasmid Elimination as a Potential Counter-Evolutionary Tactic," *Drug Discovery Today* 9 (2004), 465–67.
36. Johna DeNap and Paul Hergenrother, "Bacterial Death Comes Full Circle: Targeting Plasmid Replication in Drug-Resistant Bacteria," *Organic Biomolecular Chemistry* 3 (2005), 959–66.
37. Johna DeNap et al., "Combating Drug-Resistant Bacteria: Small Molecule Mimics of Plasmid Incompatibility as Antiplasmid Compounds," *Journal of the American Chemical Society* 126 (2004), 15402–4.
38. Virve Enne et al., "Evidence of Antibiotic Resistance Gene Silencing in *Escherichia coli*," *Antimicrobial Agents and Chemotherapy* 50 (2006), 3003–10.

39. In theory, the Flavr Savr tomato was supposed to fill grocery shelves with more flavorful tomatoes by allowing growers to pick them when fully ripe (instead of green) without worrying that they'd rot on the way to market. In reality, the modest increase in flavor wasn't enough to get consumers past their heebie-jeebies over genetically engineered produce.

40. D. G. White et al., "Inhibition of the Multiple Antibiotic Resistance (mar) Operon in Escherichia coli by Antisense DNA Analogues," Antimicrobial Agents and Chemotherapy 41 (1997), 2699–704.

41. C. Torres Viera et al., "Restoration of Vancomycin Susceptibility in Enterococcus faecalis by Antiresistance Determinant Gene Transfer," Antimicrobial Agents and Chemotherapy 45 (2001), 973–75.

42. Rene Sarno et al., "Inhibition of Aminoglycoside 6'-N acetyltransferase Type Ib-Mediated Amikacin Resistance by Antisense Oligodeoxynucleotides," Antimicrobial Agents and Chemotherapy 47 (2003), 3296–304.

43. WHO International Review Panel, "Impacts of Antimicrobial Growth Promoter Termination in Denmark: The WHO International Review Panel's Evaluation of the Termination of the Use of Antimicrobial Growth Promoters in Denmark," www.who.int/salmsurv/links/gssamrgrowthreportstory/en/.

44. McDonald's Global Policy on Antibiotic Use in Food Animals, June 2003, available online at www.mcdonalds.com/corp/values/socialrespons/market/antibiotics/global_policy.RowPar.0001.ContentPar.0001.ColumnPar.0003.File.tmp/antibiotics_policy.pdf.

45. Richard Martin, "How Ravenous Soviet Viruses Will Save the World," Wired, October 2003.

46. Peter Radetsky, "The Good Virus," Discover 17 (November 1996); Thomas Hausler, Viruses vs. Superbugs (New York: Macmillan Science, 2006).

47. Alexander "Sandro" Sulakvelidze, interviews by author, November 2006.

48. Richard Stone, "Stalin's Forgotten Cure," Science 298 (2002), 728–31.

49. Amy Ellis Nutt, "Germs That Fight Germs: How Killer Bacteria Have Defeated Our Last Antibiotic," Newark Star-Ledger, December 9, 2003.

50. Kevin Smeallie, correspondence to author, November 2006.

51. Anne Bruttin and Harald Brüssow, "Human Volunteers Receiving Escherichia coli Phage T4 Orally: A Safety Test of Phage Therapy," Antimicrobial Agents and Chemotherapy 49 (2005), 2874–78.

52. Thomas Broudy and Vincent Fischetti, "In Vivo Lysogenic Conversion of Tox– Streptococcus pyogenes to Tox+ with Lysogenic Streptococci or Free Phage," Infection and Immunity 71 (2003), 3782–86; Thomas Broudy et al., "Induction of Lysogenic Bacteriophage and Phage-Associated Toxin from Group A Streptococci During Coculture with Human Pharyngeal Cells," Infection and Immunity 69 (2001), 1440–43.

53. Steven Projan, "Phage-Inspired Antibiotics," Nature Biotechnology 22 (2004), 167–68.

54. "Phage Therapy Center Mexico, S.A. de C.V.," online at www.phagetherapycenter.com/.

55. Jing Liu et al., "Antimicrobial Drug Discovery Through Bacteriophage Genomics," Nature Biotechnology 22 (2004), 185–91.

56. Qi Cheng et al., "Removal of Group B Streptococci Colonize the Vagina and

Oropharynx of Mice with a Bacteriophage Lytic Eenzyme," *Antimicrobial Agents and Chemotherapy* 49 (2005), 111–17; J. M. Loeffler et al., "Rapid Killing of *Streptococcus pneumoniae* with a Bacteriophage Cell Wall Hydrolase," *Science* 294 (2001), 2170–72; R. Schuch et al., "A Bacteriolytic Agent That Detects and Kills *Bacillus anthracis*," *Nature* 418 (2002), 884–89; D. Nelson et al., "Prevention and Elimination of Upper Respiratory Colonization of Mice by Group A Streptococci by Using a Bacteriophage Lytic Enzyme," *Proceedings of the National Academy of Sciences* 98 (2001), 4107–12.

57. Vincent Fischetti, interviews by author, October–November 2006.

58. Britta Leverentz et al., "Biocontrol of *Listeria monocytogenes* on Fresh-Cut Produce by Treatment with Lytic Bacteriophages and a Bacteriocin," *Applied and Environmental Microbiology* 69 (2003), 4519–26; Britta Leverentz et al., "Examination of Bacteriophage as a Biocontrol Method for Salmonella on Fresh-Cut Fruit: A Model Study," *Journal of Food Protection* 64 (2001), 1116–21.

59. Harald Brüssow, "Phage Therapy: The *Escherichia coli* Experience," *Microbiology* 151 (2005), 2133–40; T. R. Calloway et al., "What Are We Doing About *Escherichia coli* 0157:H7 in Cattle," *Journal of Animal Science* 82 (2004), E93–E99.

60. H. Steiner et al., "Sequence and Specificity of Two Antibacterial Proteins Involved in Insect Immunity," *Nature* 292 (1981), 246–48.

61. T. Ganz et al., "Defensins, Natural Peptide Antibiotic of Human Neutrophils," *Journal of Clinical Investigation* 76 (1985), 1427–35.

62. Michael Zasloff, "Magainins, a Class of Antimicrobial Peptides from Xenopus Skin: Isolation, Characterization of Two Active Forms, and Partial cDNA Sequence of a Precursor," *Proceedings of the National Academy of Sciences* 84 (1987), 5449–53.

63. "Magainin, Shield Against Disease," Editorial, *New York Times*, August 9, 1987, E24; see also Lawrence Altman, "Staying Ahead of Microbes: New Progress," *New York Times*, August 4, 1987, C3.

64. Michael Zasloff, "Antimicrobial Peptides of Multicellular Organisms," *Nature* 415 (2002), 389–95.

65. Y. Ge et al., "In Vitro Antibacterial Properties of Pexiganan, an Analog of Magainin," *Antimicrobial Agents and Chemotherapy* 43 (1999), 782–88.

66. M. J. Goldman et al., "Human Beta-defensin-1 Is a Salt-Sensitive Antibiotic in Lung That Is Inactivated in Cystic Fibrosis," *Cell* 88 (1997), 553–60.

67. L. Jacob and M. Zasloff, "Potential Therapeutic Applications of Magainins and Other Antimicrobial Agents of Animal Origin," *Ciba Foundation Symposia* 186 (1994),197–23; Y. Ge et al., "In Vitro Susceptibility to Pexiganan of Bacteria Isolated from Infected Diabetic Foot Ulcers," *Diagnostic Microbiology and Infectious Disease* 35 (1999), 45–53.

68. Department of Health and Human Services, Food and Drug Administration, Center for Drug Evaluation and Research, Anti-infective Drugs Advisory Committee, 66th meeting, March 4, 1999, transcript.

69. Rexford Ahima et al., "Appetite Suppression and Weight Reduction by a Centrally Active Aminosterol," *Diabetes* 51 (2002), 2099–104.

70. Graham Bell and Pierre-Henri Gouyon, "Arming the Enemy: The Evolution of Resistance to Self-Proteins," *Microbiology* 149 (2003), 1367–75.

71. Jack Lucentini, "Antibiotic Arms Race Heats Up," *Scientist*, September 8, 2003, 29.

72. Gabriel Perron, Michael Zasloff, and Graham Bell, "Experimental Evolution of Resistance to an Antimicrobial Peptide," *Proceedings of the Royal Society: B (biology)* 273 (2006), 251–56.

73. Charlotte Schubert, "Microbes Overcome Natural Antibiotic," news@nature.com, November 2, 2005, available online at www.nature.com/news/2005/051031-5.html.

74. Rubhana Raqib et al., "Improved Outcome in Shigellosis Associated with Butyrate Induction of an Endogenous Peptide Antibiotic," *Proceedings of the National Academy of Sciences* 103 (2006), 9178–83.

75. Philip Liu et al., "Toll-like Receptor Triggering of a Vitamin D–Mediated Human Antimicrobial Response," *Science* 311 (2006), 1170–773.

76. Emma Harris, "Extreme TB Strain Threatens HIV Victims Worldwide," *Nature* 443 (2006), 131.

PART VI: BEYOND LETHAL FORCE – DEFANG, DEFLECT, AND DEPLOY

Part opener: Theodor Rosebury, *Microorganisms Indigenous to Man* (New York: McGraw-Hill, 1962) 352–53.

1. Victor Nizet, interviews by the author, November 2006.

2. Jesse Wright et al., "The Agr Radiation: An Early Event in the Evolution of Staphylococci," *Journal of Bacteriology* 187 (2005), 5585–94.

3. Jesse Wright, Rhuzong Jim, and Richard Novick, "Transient Interference with Staphylococcal Quorum Sensing Blocks Abscess Formation," *Proceedings of the National Academy of Sciences* 102 (2005), 1691–96.

4. K. A. Davis, "Ventilator Associated Pneumonia: A Review," *Journal of Intensive Care Medicine* 21 (2006), 211–26.

5. Richard Novick, interviews by the author, November 2006.

6. Phillip Coburn et al., "*Enterococcus faecalis* Senses Target Cells and in Response Expresses Cytolysin," *Science* 306 (2004), 2270–72.

7. Michael Gilmore, interview by the author, November 2006.

8. George Liu et al., "Sword and Shield: Linked Group B Streptococcal B-hemolysin/cytolysin and Carotenoid Pigment Function to Subvert Host Phagocyte Defense," *Proceedings of the National Academy of Sciences* 101 (2004), 14491–96.

9. Vivekanand Datta et al., "Mutational Analysis of the Group A Streptococcal Operon Encoding Streptolysin S and Its Virulence Role in Invasive Infection," *Molecular Microbiology* 56 (2005), 681–95; John Buchanan et al., "Dnase Expression Allows the Pathogen Group A Streptococcus to Escape Killing in Neutrophil Extracellular Traps," *Current Biology* 16 (2006), 396–400.

10. Thomas Louie et al., "Tolevamer, a Novel Nonantibiotic Polymer, Compared with Vancomycin in the Treatment of Mild to Moderately Severe *Clostridium difficile*–Associated Diarrhea," *Clinical Infectious Diseases* 43 (2006), 411–20.

11. David Davidson, senior medical director, Genzyme Corp., communication to the author, November 2006.

12. Critics of virulence-targeting vaccines have produced mathematical models that suggest that they may actually lead to a rise in virulence because they don't "pun-

ish" virulent strains with the death of their hosts. But real-world experience with virulence-targeting vaccines such as diphtheria have shown just the opposite. See Benoit Soubeyrand and Stanley Plotkin, "Microbial Evolution: Antitoxin Vaccines and Pathogen Virulence," *Nature* 414 (2001), 751–56.

13. "Drug-Resistant *Streptococcus pneumoniae* Disease," disease listing, National Center for Infectious Diseases/Division of Bacterial and Mycotic Diseases, October 6, 2005.

14. Moe Kyaw et al., "Effect of Introduction of the Pneumococcal Conjugate Vaccine on Drug-Resistant *Streptococcus pneumoniae*," *New England Journal of Medicine* 354 (2006), 1455–1524.

15. Henry Shinefield et al., "Use of *Staphylococcus aureus* Conjugate Vaccine in Patients Receiving Hemodialysis," *New England Journal of Medicine* 346 (2002), 491–96.

16. Yukiko Stranger-Jones et al., "Vaccine Assembly from Surface Proteins of *Staphylococcus aureus*," *Proceedings of the National Academy of Sciences* 103 (2006), 16942–47.

17. Yukiko Stranger-Jones, interviews by the author, November–December 2006.

18. D. G. Brockstedt et al., "Killed but Metabolically Active Microbes: A New Vaccine Paradigm for Eliciting Effector T-cell Responses and Protective Immunity," *Nature Medicine* 11 (2005), 853–60.

19. N. Porat et al., "Emergence of Penicillin-Nonsusceptible *Streptococcus pneumoniae* Clones Expressing Serotypes Not Present in the Antipneumococcal Conjugate Vaccine," *Journal of Infectious Diseases* 190 (2004), 2154–61.

20. Grace Lee et al., "Pertussis in Adolescents and Adults: Should We Vaccinate?" *Pediatrics* 115 (2005), 1675–84.

21. *Haemophilus Influenzae Type b (Hib) Vaccine: What You Need to Know* (Atlanta: Centers for Disease Control, 2006).

22. Heikki Peltola, "Worldwide *Haemophilus influenzae* Type b Disease at the Beginning of the 21st Century," *Clinical Microbiology Reviews* 13 (2000), 302–17.

23. Elie Metchnikoff, *The Prolongation of Life* (New York: G. P. Putnam's Sons, 1908).

24. Sherwood Gorbach, "The Discovery of *Lactobacillus GG*," *Nutrition Today* 31 (1996), 2S–4S.

25. H. L. DuPont, "Prevention of Diarrhea by the Probiotic *Lactobacillus GG*," *Journal of Pediatrics* 134 (1999), 1–2; T. Arvola et al., "Prophylactic *Lactobacillus GG* Reduces Antibiotic-Associated Diarrhea in Children with Respiratory Infections: A Randomized Study," *Pediatrics* 104 (1999), e64; E. Hilton et al., "Efficacy of *Lactobacillus GG* as a Diarrheal Preventive in Travelers," *Journal of Travel Medicine* 4 (1997), 41–43; J. A. Billier et al., "Treatment of Recurrent *Clostridium difficile* Colitis with *Lactobacillus GG*," *Pediatric Gastroenterology and Nutrition* 21 (1995), 224–26.

26. Andrew Bruce et al., "Recurrent Urethritis in Women," *Canadian Medical Association Journal* 108 (1973), 973–76; Andrew Bruce et al., "The Significance of Perineal Pathogens in Women," *Journal of Urology* 112 (1974), 808–10.

27. J. D. Sobel, "Is There a Protective Role for Vaginal Flora?" *Current Infectious Disease Reports* 1 (1999), 379–83; Marie Pirotta et al., "Effect of Lactobacillus in Preventing Post-antibiotic Vulvovaginal Candidiasis: A Randomised Controlled Trial," *British Medical Journal* 329 (2004), 548–51; T. Kontiokari et al., "Random-

ized Trial of Cranberry-Lingonberry Juice and *Lactobacillus* GG Drink for the Prevention of Urinary Tract Infections in Women," *British Medical Journal* 322 (2001), 1571–75.

28. Gregor Reid, "In Vitro Testing of *Lactobacillus acidophilus* NCFM as a Possible Probiotic for the Urogenital Tract," *International Dairy Journal* 10 (2000), 415–19.

29. R. C. Chan, A. W. Bruce, and G. Reid, "Adherence of Cervical, Vaginal and Distal Urethral Normal Microbial Flora to Human Uroepithelial Cells and the Inhibition of Adherence of Gram-negative Uropathogens by Competitive Exclusion," *Journal of Urology* 131 (1984), 596–601; Gregor Reid and Andrew Bruce, "Selection of *Lactobacillus* Strains for Urogenital Probiotic Applications," *Journal of Infectious Diseases* 183 (2001), S77–S80; Gillian Gardiner et al., "Persistence of *Lactobacillus fermentum* RC-14 and *Lactobacillus rhamnosus* GR-1 but not *L. rhamnosus* GG in the Human Vagina as Demonstrated by Randomly Amplified Polymorphic DNA," *Clinical and Diagnostic Laboratory Immunology* 9 (2002), 92–96.

30. Gregor Reid et al., "Probiotic Lactobacillus Dose Required to Restore and Maintain a Normal Vaginal Flora," *FEMS Immunology and Medical Microbiology* 32 (2001), 37–41; Gardiner et al., "Persistence of *Lactobacillus fermentum* RC-14."

31. Kingsley Anukam et al., "Clinical Study Comparing Probiotic *Lactobacillus* GR-1 and RC-14 with Metronidazole Vaginal Gel to Treat Symptomatic Bacterial Vaginosis," *Microbes and Infection* 8 (2006), 2772–76.

32. Jarrow Formulas, "Unique Probiotic Concept for Women Now Launched in USA," press release, April 4, 2006.

33. Ngo Thi Hoa et al., "Characterization of *Bacillus* Species Used for Oral Bacteriotherapy and Bacterioprophylaxis of Gastrointestinal Disorders," *Applied and Environmental Microbiology* 66 (2000), 5241–47; Gregor Reid, "Safe and Efficacious Probiotics: What Are They?" *Trends in Microbiology* 14 (2006), 348–52.

34. Kristian Roos et al., "Perianal Streptococcal Dermatitis. The Possible Protective Role of Alpha-streptococci Against Spread and Recurrence of Group A Streptococcal Throat Infection," *Scandinavian Journal of Primary Health Care* 17 (1999), 46–48.

35. Kristian Roos et al., "Recolonization with Selected Alpha-streptococci for Prophylaxis of Recurrent Streptococcal Pharyngotonsillitis—A Randomized Placebo-Controlled Multicentre Study," *Scandinavian Journal of Infectious Disease* 28 (1996), 459–62.

36. G. Falck et al., "Tolerance and Efficacy of Interfering Alpha-Streptococci in Recurrence of Streptococcal Pharyngotonsillitis," *Acta Oto-Laryngologica* 119 (1999), 944–48.

37. Kristian Roos et al., "Effect of Recolonisation with Interfering Alpha Streptococci on Recurrences of Acute and Secretory Otitis Media in Children: Randomised Placebo Controlled Trial," *British Medical Journal* 322 (2001), 1–4.

38. Janet Raloff, "'Bug' Spray Cuts Risk of Ear Infection," *Science News*, February 3, 2001; Lee Bowman, "Study Suggests Alternative to Antibiotics for Ear Infections," Scripps Howard News Service, January 2001; Mary Ann Moon, "Microbial Nasal Spray Wards Off Recurrent Otitis," *Family Practice News*, May 1, 2001; "A Nasal Spray to Prevent Otitis Media," *Hearing Journal*, April 1, 2001.

39. Glenn Takata et al., "Evidence Assessment of Management of Acute Otitis Media: I. The Role of Antibiotics in Treatment of Uncomplicated Acute Otitis Media," *Pediatrics* 108 (2001), 239–47.

40. Kelly Karpa, interviews by author. Karpa is a professor of pharmacology, Pennsylvania State University College of Medicine. *Bacteria for Breakfast* (Trafford Publishing 2006) is the story of Karpa's 2001 ordeal getting treatment for her six-year-old son, critically ill from recurrent *C. difficile* colitis.

41. L. V. McFarland, "Meta-analysis of Probiotics for the Prevention of Antibiotic-Associated Diarrhea and the Treatment of *Clostridium difficile* Disease," *American Journal of Gastroenterology* 101 (2006), 812–22; Mario Guslandi, "Are Probiotics Effective for Treating *Clostridium difficile* Disease and Antibiotic-Associated Diarrhea," *Nature Clinical Practice* 3 (2006), 606–7.

42. "ViroPharma Licenses Rights to Develop Novel Therapeutic for Treatment of *Clostridium difficile*," company press release, February 27, 2006.

43. See Part 4, pages 108–9.

44. R. W. Steele, "Recurrent Staphylococcal Infection in Families," *Archives of Dermatology* 116 (1980), 189–90.

45. Richard Hull, interviews by and email correspondence to the author, January 2005–November 2006.

46. Peter Andersson et al., "Persistence of *Escherichia coli* Bacteriuria Is Not Determined by Bacterial Adherence," *Infection and Immunity* 59 (1991), 2915–21; Richard Hull et al., "Virulence Properties of *Escherichia coli* 83972, a Prototype Strain Associated with Asymptomatic Bacteriuria," *Infection and Immunity* 67 (1999), 429–32.

47. Richard Hull et al., "Urinary Tract Infection Prophylaxis Using *Escherichia coli* 83972 in Spinal Cord Injured Patients," *Journal of Urology* 163 (2000), 872–77; Rabih Darouiche et al., "Pilot Trial of Bacterial Interference for Preventing Urinary Tract Infection," *Urology* 58 (2001), 339–44.

48. B. W. Trautner et al., "*Escherichia coli* 83972 Inhibits Catheter Adherence by a Broad Spectrum of Uropathogens," *Urology* 61 (2005), 1059–62; Rabih Darouiche et al., "Bacterial Interference for Prevention of Urinary Tract Infection: A Prospective, Randomized Placebo-Controlled, Double-Blind Pilot Trial," *Clinical Infectious Diseases* 41 (2005), 1531–34. A larger, longer trial is running through 2008.

49. Hull et al., "Virulence Properties of *Escherichia coli* 83972"; Richard Hull et al., "Role of Type 1 Fimbria- and P Fimbria-Specific Adherence in Colonization of the Neurogenic Human Bladder by *Escherichia coli*," *Infection and Immunity* 70 (2002), 6481–84.

50. Jeffrey Hillman, "Lactate Dehydrogenase Mutants of *Streptococcus mutans*: Isolation and Preliminary Characterization," *Infection and Immunity* 21 (1978), 206–12.

51. R. J. Berkowitz and P. Jones, "Mouth-to-Mouth Transmission of the Bacterium *Streptococcus mutans* Between Mother and Child," *Archives of Oral Biology* 30 (1985), 377–79.

52. Jeffrey Hillman et al., "Isolation of a *Streptococcus mutans* Strain Producing a Novel Bacteriocin," *Infection and Immunity* 44 (1984), 141–44.

53. Jeffrey Hillman et al., "Colonization of the Human Oral Cavity by a Strain of *Streptococcus mutans* Mutant Producing Increased Bacteriocin," *Journal of Dental Research* 66 (1987), 1092–94.

54. Jeffrey Hillman et al., "Construction and Characterization of an Effector Strain of *Streptococcus mutans* for Replacement Therapy of Dental Caries," *Infection and Immunity* 68 (2000), 543–49.

55. Maikel Peppelenbosch, interview by the author, February 1, 2007.

56. Lothar Steidler et al., "Treatment of Murine Colitis by *Lactococcus lactis* Secreting Interleukin-10," *Science* 5483 (2000), 1352–55.

57. Henri Braat et al., "A Phase I Trial with Transgenic Bacteria Expressing Interleukin-10 in Crohn's Disease," *Clinical Gastroenterology and Hepatology* 4 (2006), 754–59.

58. Janice Liu et al., "Activity of HIV Entry and Fusion Inhibitors Expressed by the Human Vaginal Colonizing Probiotic *Lactobacillus reuteri* PC-14," *Cellular Microbiology* 9 (2007), 120–30.

59. Peter Lee, interview by and email correspondence to the author, January 2005–December 2006.

60. N. Sewankambo et al., "HIV-1 Infection Associated with Abnormal Vaginal Flora Morphology and Bacterial Vaginosis," *Lancet* 350 (1997), 546–50; H. L. Martin et al., "Vaginal Lactobacilli, Microbial Flora, and the Risk of Human Immunodeficiency Virus Type 1 and Sexually Transmitted Disease Acquisition," *Journal of Infectious Diseases* 180 (1999), 1863–68.

61. Theresa L.-Y. Chang, "Inhibition of HIV Infectivity by a Natural Human Isolate of *Lactobacillus jensenii* Engineered to Express Functional Two-Domain CD4," *Proceedings of the National Academy of Sciences* 100 (2003), 11672–77.

62. M. K. Boyd et al., "Discovery of Cyanovirin-N, a Novel Human Immunodeficiency Virus-Inactivating Protein That Binds Viral Surface Envelop Glycoprotein gh120: Potential Applications to Microbicide Development," *Antimicrobial Agents and Chemotherapy* 41 (1997), 1521–30; C.C.P. Tsai et al., "Cyanovirin-N Inhibits AIDS Virus Infections in Vaginal Transmission Models," *AIDS Research and Human Retroviruses* 20 (2004), 11–18.

63. Xiaowen Liu et al., "Engineered Vaginal *Lactobacillus* Strain for Mucosal Delivery of the Human Immunodeficiency Virus Inhibitor Cyanovirin-N," *Antimicrobial Agents and Chemotherapy* 50 (2006), 3250–59.

64. D. Medaglini et al., "Mucosal and Systemic Immune Responses to a Recombinant Protein Expressed on the Surface of the Oral Commensal Bacterium *Streptococcus gordonii* After Oral Colonization," *Proceedings of the National Academy of Sciences* 92 (1995), 6868–72.

65. Ashu Sharma et al., "Oral Immunization with Recombinant *Streptococcus gordonii* Expressing *Porphyromonas gingivalis* FimA Domains," *Infection and Immunity* 69 (2001), 2928–34.

66. Corinne Grangette et al., "Protection Against Tetanus Toxin After Intragastric Administration of Two Recombinant Lactic Acid Bacteria: Impact of Strain Viability and In Vitro Persistence," *Vaccine* 20 (2002), 3304–9; B. Corthesy et al., "Oral Immunization of Mice with Lactic Acid Bacteria Producing *Helicobacter pylori* Urease B Subunit Partially Protects Against Challenge with *Helicobacter felis*," *Journal of Infectious Diseases* 192 (2005), 1441–49.

67. L. Scheppler et al., "Intranasal Immunisation Using Recombinant *Lactobacillus johnsonii* as a New Strategy to Prevent Allergic Disease," *Vaccine* 9 (2005), 1126–34.
68. Scheppler et al., "Intranasal Immunisation Using Recombinant *Lactobacillus johnsonii*"; B. Stadler et al., "Lactic Acid Bacteria as Agents for Preventing Allergy," U.S. Patent Application 200402655290.
69. Eva Medina and Carlos Alberto Guzman, "Use of Live Bacterial Vaccine Vectors for Antigen Delivery: Potential and Limitations," *Vaccine* 19 (2001), 1573–80.
70. Joe Cummins, e-mail correspondence with the author, November 2006.
71. Ronald Jackson et al., "Expression of Mouse Interleukin-4 by Recombinant Ectromelia Virus Suppresses Cytolytic Lymphocyte Responses and Overcomes Genetic Resistance in Mousepox," *Journal of Virology* 75 (2001), 1205–10; Rachel Nowak, "Disaster in the Making," *New Scientist*, January 3, 2001, 4–5.
72. Freedom of Information Summary NADA 141-101 (FDA approval of Preempt, with summary of indications for use and effectiveness), www.fda.gov/cvm/FOI/886.htm.
73. J. Raloff, "Spray Guards Chicks from Infections," *Science News*, March 28, 1998, 196.
74. Todd Callaway, interviews by the author, September–December 2006.
75. Kenneth Genovese et al., "Protection of Suckling Neonatal Pigs Against Infection with an Enterotoxigenic *Escherichia coli* Expressing 987P Fimbriae by the Administration of a Bacterial Competitive Exclusion Culture," *Microbial Ecology in Health and Disease* 13 (2001), 223–28; Roger Harvey et al., "Use of Competitive Exclusion to Control Enterotoxigenic Strains of *Escherichia coli* Weaned Pigs," *Journal of Animal Science* 83 (2005), E44–E47.
76. R. E. Holland, "Infectious Causes of Diarrhea in Young Farm Animals," *Clinical Microbiology Reviews* 3 (1990), 345–75; S. Tzipori, "The Relative Importance of Enteric Pathogens Affecting Neonates of Domestic Animals," *Advances in Veterinary Science and Comparative Medicine* 29 (1985), 103–206.
77. Francis Forst, interview by the author, November 28, 2006.
78. David Thaler, interviews by and email correspondence to the author, December 2004–December 2006, and David Thaler, "The Microbial Neolithic Revolution," (unpublished).
79. David Thaler, "The Evolution of Genetic Intelligence," *Science* 264 (1994), 224–25; David Thaler, "Hereditary Stability and Variation in Evolution and Development," *Evolution and Development* 1 (1999), 113–22.

PART VII: FIXING THE PATIENT
Part opener: Lewis Thomas, *Lives of a Cell: Notes of a Biology Watcher* (New York: Penguin, 1978), 78.

1. Kevin Tracey, interviews by the author, November 2006. For more about Janice's story and Tracey's pioneering efforts to understand septic shock and severe sepsis, I highly recommend his evocative and deeply informative book *Fatal Sequence: The Killer Within* (Washington, D.C.: Dana Press, 2005).

2. Derek Angus et al., "Epidemiology of Severe Sepsis in the United States: Analysis of Incidence, Outcome, and Associated Costs of Care," *Critical Care Medicine* 29 (2001), 1303–10.

3. Brian Beutler et al., "Passive Immunization Against Cachectin/Tumor Necrosis Factor Protects Mice from Lethal Effect of Endotoxin," *Science* 229 (1985), 869–71.

4. Kevin Tracey et al., "Shock and Tissue Injury Induced by Recombinant Human Cachectin," *Science* 234 (1986), 470–74.

5. Kevin Tracey et al., "Anti-cachectin/TNF Monoclonal Antibodies Prevent Septic Shock During Lethal Bacteraemia," *Nature* 330 (1987), 662–64.

6. Huan Wang et al., "HMG-1 as a Late Mediator of Endotoxin Lethality in Mice," *Science* 285 (1999), 248–51; Huan Yang et al., "Reversing Established Sepsis with Antagonists of Endogenous High-Mobility Group Box 1," *Proceedings of the National Academy of Sciences* 101 (2004), 296–301.

7. L. V. Borovikova et al., "Vagus Nerve Stimulation Attenuates the Systemic Inflammatory Response to Endotoxin," *Nature* 405 (2000), 458–62.

8. Hong Wang et al., "Nicotinic Acetylcholine Receptor alpha7 Subunit Is an Essential Regulator of Inflammation," *Nature* 421 (2003), 384–88; Kevin Tracey, "The Inflammatory Reflex," *Nature* 420 (2002), 853–59.

9. Thomas Bernik et al., "Pharmacological Stimulation of the Cholinergic Anti-inflammatory Pathway," *Journal of Experimental Medicine* 195 (2002), 781–88; H. Wang et al., "Cholinergic Agonists Inhibit HMGB1 Release and Improve Survival in Experimental Sepsis," *Nature Medicine* 10 (2004), 1216–21.

10. Greta Van den Berghe et al., "Intensive Insulin Therapy in Critically Ill Patients," *New England Journal of Medicine* 345 (2001), 1359–67.

11. Greg Martin et al., "The Epidemiology of Sepsis in the United States from 1979 Through 2000," *New England Journal of Medicine* 348 (2003), 1546–54.

12. Derek Angus, "Sepsis on the Rise in the United States," presentation to the 32nd annual Critical Care Congress, San Antonio, Texas, 2003; Derek Angus et al., "Epidemiology of Severe Sepsis in the United States: Analysis of Incidence, Outcome, and Associated Costs of Care," *Critical Care Medicine* 29 (2001), 1303–10.

13. Garth Ehrlich et al., "Device-Related Infections of Prosthetic Devices in the United States, Table 1 in Engineering Approaches for the Detection and Control of Orthopaedic Biofilm Infections," *Clinical Orthopaedics and Related Research* 2005 (437), 59–66. Note: The referenced table omits ocular lens implants, of which 1 to 2 million are implanted each year in the United States, with an infection rate of 7 to 10 percent. Source: A. Hornblass et al., "Current Techniques of Enucleation: A Survey of 5,439 Intraorbital Implants and a Review of the Literature," *Ophthalmic Plastic and Reconstructive Surgery* 11 (1995), 77–86.

14. Garth Ehrlich et al., "Intelligent Implants to Battle Biofilms," *ASM News* 70 (2004), 127–33.

15. Joel Epstein et al., "A Survey of Antibiotic Use in Dentistry," *Journal of the American Dental Association* 131 (2000), 1600–1609; G. W. Meyer and A. L. Artis, "Antibiotic Prophylaxis for Orthopedic Prostheses and GI Procedures: A Report of a Survey," *American Journal of Gastroenterology* 92 (1997), 989–91; P. B. Lockhart et al., "Decision-Making on the Use of Antimicrobial Prophylaxis for Dental Proce-

dures: A Survey of Infectious Disease Consultants and a Review," *Clinical Infectious Diseases* 34 (2002), 1621–26.

16. Lucas Hoffman et al., "Aminoglycoside Antibiotics Induce Bacterial Biofilm Formation," *Nature* 436 (2005), 1171–75.

17. C. von Eiff et al., "Modern Strategies in the Prevention of Implant-Associated Infections," *International Journal of Artificial Organs* 28 (2005), 1146–56.

18. D. Neut et al., "Residual Gentamicin-Release from Antibiotic-Loaded Polymethylmethacrylate Beads After Five Years of Implantation," *Biomaterials* 24 (2003), 1829–31.

19. R. G. Seipelt et al., "The St. Jude 'Silzone' Valve: Midterm Results in Treatment of Active Endocarditis," *Annals of Thoracic Surgery* 72 (2001), 758–62, esp. 762–63.

20. G. Cook et al., "Direct Confocal Microscopy Studies of the Bacterial Colonization In Vitro of a Silver-Coated Heart Valve Sewing Cuff," *International Journal of Antimicrobial Agents* 13 (2000), 169–73.

21. Michael Surette et al., "Quorum Sensing in *Escherichia coli*, *Salmonella typhimurium*, and *Vibrio harveyi*: A New Family of Genes Responsible for Autoinducer Production," *Proceedings of the National Academies of Science* 96 (1999), 1639–44; Stephan Schauder et al., "The luxS Family of Bacterial Autoinducers: Biosynthesis of a Novel Quorum-Sensing Signal Molecule," *Molecular Microbiology* 41 (2001), 463–76.

22. Bonnie Bassler, interviews by the author, November–December 2005.

23. Karina Xavier and Bonnie Bassler, "Interference with AI-2-Mediated Bacterial Cell-Cell Communication," *Nature* 437 (2005), 750–53.

24. E. Barth et al., "In Vitro and In Vivo Comparative Colonization of *Staphylococcus aureus* and *Staphylococcus epidermidis* on Orthopaedic Implant Materials," *Biomaterials* 10 (1989), 325–28.

25. Andrea Giacometti et al., "RNA III Inhibiting Peptide Inhibits In Vivo Biofilm Formation by Drug-Resistant *Staphylococcus aureus*," *Antimicrobial Agents and Chemotherapy* 47 (2003), 1979–83.

26. Garth Ehrlich et al., "Engineering Approaches for the Detection and Control of Orthopaedic Biofilm Infections," *Clinical Orthopaedics and Related Research* 437 (2005), 59–66.

27. Philip Stewart and Ross Carlson, "Anti-biofilm Properties of Chitosan-Coated Surfaces," 232nd national meeting of the American Chemical Society, poster presentation COLL 021, September 10, 2006.

28. Anneta Razatos et al., "Force Measurements Between Bacteria and Polyethylene Glycol Coated Surfaces," *Langmuir* 16 (2000), 9155–58; David Adam, "Bacteria Get the Brush Off," www.nature.com/nsu/001221/0012216.html.

29. MedImmune, 2005 Annual Report, Gaithersburg, MD.

30. A. P. Simopoulos, "Omega-3 Fatty Acids in Inflammation and Autoimmune Diseases," *Journal of the American College of Nutrition* 21 (2002), 495–505; F. Holguin et al., "Cardiac Autonomic Changes Associated with Fish Oil vs Soy Oil Supplmentation in the Elderly," *Chest* 127 (2005), 1102–07; A. A. Berbert et al., "Supplementation of Fish Oil and Olive Oil in Patients with Rheumatoid Arthritis," *Nutrition* 21 (2005), 131–36.

31. Misha Luyer et al., "Nutritional Stimulation of Cholecystokinin Receptors In-

hibits Inflammation via the Vagus Nerve," *Journal of Experimental Medicine* 202 (2005), 1023–29.

32. T. Chen et al., "Mononuclear Cell Response to Enterobacteria and Gram-positive Cell Walls of Normal Intestinal Microbiota in Early Rheumatoid Arthritis and Other Inflammatory Arthritides," *Clinical and Experimental Rheumatology* 20 (2002), 193–200; Erika Isolauri, "Probiotics in Human Disease," *American Journal of Clinical Nutrition* 73 (2001), 1142S–1146S; Kent Erickson and Neil Hubbard, "Probiotic Immunomodulation in Health and Disease," *Journal of Nutrition* 130 (2000), 403S–409S.

33. K. Hatakka et al., "Effects of Probiotic Therapy on the Activity and Activation of Mild Rheumatoid Arthritis—A Pilot Study," *Scandinavian Journal of Rheumatology* 32 (2003), 211–15; Ehud Baharav et al., "*Lactobacillus* GG Bacteria Ameliorate Arthritis in Lewis Rats," *Journal of Nutrition* 134 (2004), 1964–69; T. M. Chapman et al., "VSL#3 Probiotic Mixture: A Review of Its Use in Chronic Inflammatory Bowel Diseases," *Drugs* 66 (2006), 1371–87; O. Karimi et al., "Probiotics (VSL#3) in Arthralgia in Patients with Ulcerative Colitis and Crohn's Disease: A Pilot Study," *Drugs Today* 41 (2005), 453–59; Philippe Marteau et al., "Protection from Gastrointestinal Diseases with the Use of Probiotics," *American Journal of Clinical Nutrition* 73 (2001), 430S–436S.

34. Henri Braat et al., "A Phase 1 Trial with Transgenic Bacteria Expressing Interleukin-10 in Crohn's Disease," *Clinical Gastroenterology and Hepatology* 4 (2006), 754–59.

35. M. Kalliomaki et al., "Probiotics in Primary Prevention of Atopic Disease: A Randomised Placebo-Controlled Trial," *Lancet* 357 (2001), 1076–79; M. Kalliomaki et al., "Probiotics and Prevention of Atopic Disease: 4-year Follow-up of a Randomised Placebo-Controlled Trial," *Lancet* 361 (2003), 1869–71.

36. M. Viljanen, "Probiotics in the Treatment of Atopic Eczema/Dermatitis Syndrome in Infants: A Double-Blind Placebo-Controlled Trial," *Allergy* 60 (2005), 494–500.

37. F. Calcinaro et al., "Oral Probiotic Administration Induces Interleukin-10 Production and Prevents Spontaneous Autoimmune Diabetes in the Non-obese Diabetic Mouse," *Diabetologia* 48 (2005), 1565–75; Mihoko Tabuchi et al., "Antidiabetic Effect of *Lactobacillus* GG in Streptozotocin-Induced Diabetic Rats," *Bioscience, Biotechnology and Biochemistry* 67 (2003), 1421–24; T. Matsuzaki et al., "Prevention of Onset in an Insulin-Dependent Diabetes Mellitus Model, NOD Mice, by Oral Feeding of *Lactobacillus casei*," *Acta Pathologica, Microbiologica et Immunologica Scandinavica* 105 (1997), 643–49; T. Matsuzaki, "Antidiabetic Effects of an Oral Administration of *Lactobacillus casei* in a Non-insulin-dependent Diabetes Mellitus Model Using KK-Ay Mice," *Endocrinology Journal* 44 (1997), 357–65.

38. M. Ljungberg et al., "Probiotics for the Prevention of Beta Cell Autoimmunity in Children at Genetic Risk of Type 1 Diabetes—The PRODIA Study," *Annals of the New York Academy of Sciences* 1079 (2006), 360–64.

39. Martin Bachmann, interview by author, October 16, 2006; "Cytos Biotechnology Updates on Development of Allergy Vaccine," company press release, June 12, 2006; Cytos Biotechnology, 2006 Third Quarter Report, September 30, 2006.

40. Cytos Biotechnology, "Vaccine Candidate CYT005-AllQbG10 for Allergic Diseases Shows Significant Efficacy in Phase 11a Study," press release, December 14,

2005; Cytos Biotechnology, "Vaccine to Treat Allergic Diseases Shows Significant Long-term Efficacy in House Dust Mite Allergy Patients," press release, April 25, 2006; Cytos Biotechnology, 2006 Third Quarter Report, September 30, 2006.

41. Martin Bachmann et al., "The Influence of Antigen Organization on B Cell Responsiveness," *Science* 262 (1993), 1448–51.

42. Tazio Storni et al., "Nonmethylated CG Motifs Packaged into Virus-like Particles Induce Protective Cytotoxic T Cell Responses in the Absence of Systemic Side Effects," *Journal of Immunology* 172 (2004), 1777–85.

43. Y. Sato et al., "Immunostimulatory DNA Sequences Necessary for Effective Intradermal Gene Immunization," *Science* 273 (1996), 352–54; M. Roman et al., "Immunostimulatory DNA Sequences Function as T Helper-1-Promoting Adjuvants," *Nature Medicine* 3 (1997), 849–54.

44. David Broide et al., "Immunostimulatory DNA Sequences Inhibit IL-5, Eosinophilic Inflammation and Airway Hyperresponsiveness in Mice," *Journal of Immunology* 161 (1998), 7054–62; David Broide et al., "Systemic Administration of Immunostimulatory DNA Sequences Mediates Reversible Inhibition of TH2 Responses in a Mouse Model of Asthma," *Journal of Clinical Immunology* 21 (2001), 175–82; Arash Ronaghy et al., "Immunostimulatory DNA Sequences Influence the Course of Adjuvant Arthritis," *Journal of Immunology* 168 (2002), 51–56; Daniel Rachmilewitz et al., "Immunostimulatory DNA Ameliorates Experimental and Spontaneous Murine Colitis," *Gastroenterology* 122 (2002), 1428–41; Omar Duramad et al., "Inhibitors of TLR-9 Act on Multiple Cell Subsets in Mouse and Man *In Vitro* and Prevent Death *In Vivo* from Systemic Inflammation," *Journal of Immunology* 174 (2004), 5193–5200; Franck Barrat et al., "Nucleic Acids of Mammalian Origin Can Act as Endogenous Ligands for Toll-like Receptors and May Promote Systemic Lupus Erythematosus," *Journal of Experimental Microbiology* 202 (2005), 1131–39.

45. Peter Creticos et al., "Immunotherapy with a Ragweed-toll-like Receptor 9 Agonist Vaccine for Allergic Rhinitis," *New England Journal of Medicine* 355 (2006), 1445–55.

CODA: EMBRACING THE MICROBIOME

1. Joshua Lederberg, interview by and email correspondence to the author, December 4, 2004.

FURTHER READING

For those wanting to delve further into the world of bacteria and the challenges of living in their world, I highly recommend the following books and reports.

Buckley, Merry. *The Genomics of Disease-Causing Organisms: Mapping a Strategy for Discovery and Defense*. This July 2004 American Academy of Microbiology colloquia report is available online at www.asm.org/Academy/index.asp?bid=29532.

Hart, Tony. *Microterrors*. Richmond Hill, Ont.: Firefly Books, 2004. A field guide to disease-causing bacteria, viruses, fungi, and protists, gorgeously illustrated with colorized micrographs.

Levy, Stuart.*The Antibiotic Paradox*, 2nd ed. New York: Perseus, 2002. An updated edition of the classic text that woke the world up to the danger of antibiotic resistance.

Margulis, Lynn. *Symbiotic Planet*. New York: Basic Books, 1998.

Margulis, Lynn, and Dorion Sagan. *Acquiring Genomes: A Theory of the Origins of Species*. New York: Basic Books, 2002. These two books cover Margulis's groundbreaking theories of endosymbiosis and her controversial ideas of how species originate from the wholesale acquisition of new, heritable traits.

Salyers, Abigail, and Dixie Whitt. *Microbiology: Diversity, Disease, and the Environment*. Bethesda, Md.: Fitzgerald Science Press, 2001. A college textbook accessible enough for the armchair scientist.

Salyers, Abigail, and Dixie Whitt. *Revenge of the Microbes*. Washington, D.C.: ASM Press, 2005. More on how specific antibiotics work and why they increasingly don't.

Walker, Richard, and Merry Buckley. *Probiotic Microbes: The Scientific Basis*. This June 2006 American Academy of Microbiology colloquia report is available online at www.asm.org/Academy/index.asp?bid=43351.

Wilson, Michael. *Microbial Inhabitants of Humans*. New York: Cambridge University Press, 2006. The definitive textbook on what lives where and does what on planet human.

ACKNOWLEDGMENTS

While every book is a collaboration, this is particularly true of the collection of words in your hands. I can only begin to thank the scores of scientists, reference librarians, and colleagues who graciously submitted to my questions, requests for information, and early drafts. This book was also made possible by several generous grants and fellowships.

A directors grant from the Alfred P. Sloan Foundation enabled me to deepen and rigorously fact-check my research, as well as bring on board an impressive team of scientific advisers. First and foremost of these has been Abigail Salyers, professor of microbiology at the University of Illinois, former president of the American Society for Microbiology, and a prolific and engaging author and science communicator. In addition to Abigail's bow-to-stern advisement, several chapters of this book have benefited from the vetting of Harvard immunologist Dale Umetsu, Columbia University microbiologist Aaron Mitchell, and pathologist Frank Lowy, of Columbia's College of Physicians and Surgeons. Special thanks to Sloan Foundation program director Doron Weber, who saw this project's potential.

Thanks go also to the Fund for Investigative Journalism for the tremendous boost of its 2005 Book Prize, and to the Alicia Patterson Foundation for the 2005 fellowship that got this book off the ground with a series of feature stories for the *APF Reporter*. FIJ executive director John Hyde and APF executive director Margaret Engel infused their organizations' financial support with their personal encouragement.

On the literary side, my work has greatly benefited from a squad of mentors and editors. The book proposal was born during a career-invigorating five months in Sam Freedman's rightly famous book-writing seminar at Columbia Journalism School. My agent, Regula Noetzli, ensured that the proposal reached the right hands: those of Joe Wisnovsky, my able editor at Hill and Wang and a wise soul whose perspective has kept me sane these last two years. Marguerite Holloway, my master's thesis adviser at Columbia, went far beyond her professorial obligations to help shape several core chapters of this book. Longtime *Parenting* magazine editor Maura Rhodes bravely launched her freelance editing career by helping me with this daunting project. And I'm forever in the debt of my two wonderful sisters-in-law, Sharin Sachs and Cathy Snyder, both of whom improved this book with their careful reading and feedback.

It is no understatement to say that this book would not have been possible without the research assistance of Neil Silvera of Columbia's Biological Sciences Library. Thanks, Neil!

Special thanks also to evolutionary cell biologist Lynn Margulis, of the University of Massachusetts, Amherst, whose too-brief mention in this book belies her importance to our understanding of all cells prokaryotic and eukaryotic. I will forever cherish the time I spent with Lynn during the researching of this book.

Last but never least, my husband, Gary, and daughter, Eva, sustained this effort with their love, patience, and unflagging moral support.

INDEX

Academic Medical Center (Amsterdam), 208
acetylcholine, 223–24
acid reflux, 53, 54
acinetobacter, 165
acne, antibiotics for, 155, 156
actinomyces, 39, 240*n12*
Actinomycetales, 66
adaptive immunity, 89–91, 100
Addison's disease, 82
adhesins, 117
aerobic bacteria, 39, 42
Agriculture, U.S. Department of, 49, 144; Agricultural Research Service (ARS), 176, 212–13, 216; Food Safety and Inspection Service, 216
AIDS, *see* HIV/AIDS
air pollution, 85
Aldrin, Buzz, 48
allergies, 8, 10, 73–86, 91, 232–34; depression and, 102; hygiene hypothesis of, 79–81, 85, 87; and "old friends" theory of immune response, 99–100; probiotics for, 231; vaccines against, 88–89, 93, 95, 97, 98, 232–34
Allgemeine Krankenhaus (Vienna), 22
Al-Razi, 76
Alzheimer's disease, 12, 69
Amazon Basin, Stone Age tribes of, 153
American Academy of Family Physicians, 156
American Academy of Pediatrics, 155, 198
American Culture Collection, 200
American Journal of Respiratory and Critical Care Medicine, The, 85

American Medical Association, 95; journal of, see *Journal of the American Medical Association, The*
American Society for Microbiology, 49
amikacin, 168
amoxicillin, 125, 157, 265*n34*
ampicillin, 166, 167, 265*n34*
anaerobic bacteria, 38, 39, 42; intestinal, 45–51, 57, 58, 64, 120
anaphylaxis, 8, 9, 77, 88
Angus, Derek, 224–25
Animal Health Institute, 142, 145
anthrax, 24, 29; lysins for use against, 175; vaccine against, 25
antibiotic-resistant bacteria, 9–10, 12, 94, 105–148; agricultural practices and, 141–48, 168–70; antimicrobial peptides and, 178; biofilms of, 225–26; community-acquired, 125–31; gene-swapping methods of, 111–16; hospital outbreaks of, 115–24; in intestinal microflora, 131–36; medical practices promoting, 155, 157, 159–61; probiotics and, 196, 198; in soil, 137–40; techniques for counteracting resistance, 165–68; *see also* MRSA; VRE; VRSA
antibiotics, 31–32, 39, 43, 237; agricultural, 10, 97, 141–48, 156, 168–70, 213, 214; broad-spectrum, *see* broad-spectrum antibiotics; dormancy caused by, 68; implants releasing, 226–27; intestinal bacteria killed by, 50–51, 58, 99, 189, 194; during labor, 37; natural, 204; nineteenth-century, 54; overuse of, 81, 121–22; prodrug,